T0321606

Probability and Expectation

Mathematical Olympiad Series

ISSN: 1793-8570

Published

Vol. 5 Selected Problems of the Vietnamese Olympiad (1962–2009)
by Le Hai Chau (Ministry of Education and Training, Vietnam) &
Le Hai Khoi (Nanyang Technology University, Singapore)

Vol. 6 Lecture Notes on Mathematical Olympiad Courses:
For Junior Section (In 2 Volumes)
by Xu Jiagu

Vol. 7 A Second Step to Mathematical Olympiad Problems
by Derek Holton (University of Otago, New Zealand &
University of Melbourne, Australia)

Vol. 8 Lecture Notes on Mathematical Olympiad Courses:
For Senior Section (In 2 Volumes)
by Xu Jiagu

Vol. 9 Mathemaitcal Olympiad in China (2009–2010)
edited by Bin Xiong (East China Normal University, China) &
Peng Yee Lee (Nanyang Technological University, Singapore)

Vol. 11 Methods and Techniques for Proving Inequalities
by Yong Su (Stanford University, USA) &
Bin Xiong (East China Normal University, China)

Vol. 12 Geometric Inequalities
by Gangsong Leng (Shanghai University, China)
translated by: Yongming Liu (East China Normal University, China)

Vol. 13 Combinatorial Extremization
by Yuefeng Feng (Shenzhen Senior High School, China)

Vol. 14 Probability and Expectation
by Zun Shan (Nanjing Normal University, China)
translated by: Shanping Wang (East China Normal University, China)

The complete list of the published volumes in the series can be found at
http://www.worldscientific.com/series/mos

Vol. 14 | Mathematical Olympiad Series

Probability and Expectation

Original Authors

Zun Shan *Nanjing Normal University, China*

English Translators

Shanping Wang *East China Normal University, China*

Copy Editors

Ming Ni *East China Normal University Press, China*

Lingzhi Kong *East China Normal University Press, China*

Published by

East China Normal University Press
3663 North Zhongshan Road
Shanghai 200062
China

and

World Scientific Publishing Co. Pte. Ltd.
5 Toh Tuck Link, Singapore 596224
USA office: 27 Warren Street, Suite 401-402, Hackensack, NJ 07601
UK office: 57 Shelton Street, Covent Garden, London WC2H 9HE

British Library Cataloguing-in-Publication Data
A catalogue record for this book is available from the British Library.

Mathematical Olympiad Series — Vol. 14
PROBABILITY AND EXPECTATION

ISBN 978-981-3141-48-3
ISBN 978-981-3141-49-0 (pbk)

Printed in Singapore

Introduction

Probability theory is an important branch of mathematics, with wide applications in many fields. It is not only a required course for students of science and technology at universities, but also has entered into Chinese high school textbooks now.

This little book will, in an interesting problem-solving way, explain what probability theory is: its concepts, methods and meanings; particularly, two important concepts — *probability* and *mathematical expectation* (briefly *expectation*) — are emphasized. It consists of 65 problems, appended by 107 exercises and their answers.

As an extracurricular book providing supplement materials to and advanced knowledge beyond high school textbooks, its aim is to stimulate study interests of students and broaden their knowledge horizons. Some problems were given a little deeper treatment, which can be used as topics for explorative study; and they can also be skipped temporarily if a reader feels difficult to understand them at the beginning.

It is presupposed that our readers possess a knowledge of permutations and combinations, and it would be better if they have already learned basic probability theory from their textbooks. However, in order to avoid repetition, we mention as little as possible the contents of textbooks.

It is a *random event* that this little book reaches you. I do not know how much the *probability* that this event occurs is. However, it is my *expectation* that this book could reach you, which means that you have a special affinity with it.

Contents

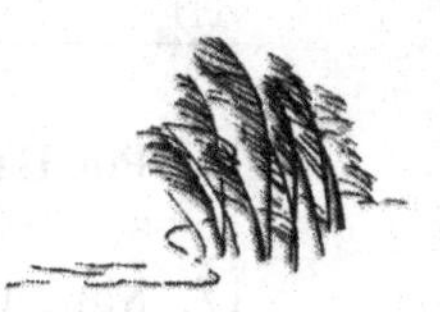

0. Basic Knowledge

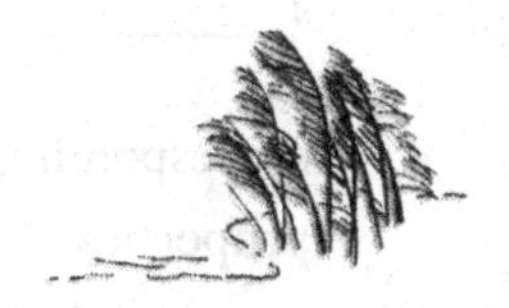

For convenient reference, some basic terms and concepts in the probability theory are listed below.

1. *Sample point*: An experiment (e.g. throwing a dice) may produce several possible results, each of them being called *a sample point*, or *an elementary event*.

2. *Elementary event*: See **1**.

3. *Sample space*: The set of all sample points is called *the sample space*, or *the total* of all elementary events, denoted by I in this book.

4. *Random event*: A subset of the sample space is called *a random event*, briefly *an event*.

5. *Event*: See **4**.

6. *Certain event*: The event that will happen certainly in an experiment is called *the certain event*, i.e. the sample space I, whose probability equals 1, i.e. $P(I) = 1$.

7. *Impossible event*: The event that cannot happen in an experiment is called the impossible event, i.e. the null set ϕ, whose probability equals 0, i.e. $P(\phi) = 0$.

8. *Mutually exclusive events*: Events A and B are called *mutually exclusive*, or *mutually inconsistent*, if their simultaneous occurrence is impossible, i.e. $A \cap B = \phi$,

9. *Mutually inconsistent events*: See **8**.

10. *Sum event*: Event $A \cup B$ is called *the sum event* of A and B.

11. *Product event*: Event $A \cap B$ is called the product event of A and B.

12. *Probability*: *The probability* is a measure defined on the sample space I, such that for every event A, there is a real number corresponding to it, denoted by $P(A)$, with the following three properties:

(1) $P(A) \geqslant 0$ (nonnegativity);

(2) $P(I) = 1$;

(3) $P(A \cup B) = P(A) + P(B)$, if A and B are mutually exclusive events (additivity).

13. *Frequency*: Suppose n experiments are carried out under the same environment, and event A happens m times among them. Then we say *the frequency* of the occurrence of A is $\frac{m}{n}$.

14. *Classical probability model*: Suppose there are n possible results, each with the same occurrence probability, in an experiment. Then we say this experiment belongs to *the classical probability model*, or *the equal probability model*, in which the occurrence probability of every result is the same $\frac{1}{n}$.

15. *Equal probability model*: See **14**.

16. *Complementary events*: Suppose events A and B satisfy

$$A \cap B = \phi \text{ and } A \cup B = I.$$

Then we say they are *complementary events*, and in this case B can be written as $\bar{A}$. For them we have a frequently used formula

$$P(\bar{A}) = 1 - P(A).$$

17. *Conditional probability*: Under the condition that event A has happened, the probability that event B happens is called *the conditional probability*, expressed as $P(B \mid A)$. Since

$$P(AB) = P(A) \cdot P(B \mid A),$$

so we have

$$P(B \mid A) = \frac{P(AB)}{P(A)}.$$

Please note the difference among $P(B \mid A)$, $P(B)$ and $P(A \mid B)$: $P(B)$ is the probability that B happens (unconditional), $P(B \mid A)$ is the probability of the occurrence of B under the condition that A has happened, and $P(A \mid B)$ is the probability of the occurrence of A under the condition that B has happened.

18. *Independent events*: If the occurrence of event A does not affect that of event B, i.e.

$$P(B \mid A) = P(B),$$

then we say that A and B are *independent events*; at this time we get easily

$$P(AB) = P(A) \cdot P(B) \text{ and, } P(A \mid B) = P(A).$$

It means that the occurrence of B also has no effect on the occurrence of A. So A and B are independent with each other.

19. *Total probability formula*: Suppose the sample space I can be divided into a set of pairwise disjoint events $B_1, B_2, \ldots, B_n$, i.e.

$$B_1 \cup B_2 \cup \cdots \cup B_n = I \text{ and, } B_i \cap B_j = \phi (1 \leqslant i < j \leqslant n).$$

Then the probability that event A happens is

$$P(A) = \sum_{i=1}^{n} P(A \mid B_i) \cdot P(B_i).$$

20. *Bayes' formula*: Let $B_1, B_2, \ldots, B_n$ be a partition of the sample space I. Under the condition that $P(B_i)$ and $P(A \mid B_i)$ $(i = 1, 2, \ldots, n)$ are known, we can use the expression

$$P(B_j \mid A) = \frac{P(AB_j)}{P(A)} = \frac{P(A \mid B_j) \cdot P(B_j)}{\sum_{i=1}^{n} P(A \mid B_i) \cdot P(B_i)}$$

to find $P(B_j \mid A)$ $(j = 1, 2, \ldots, n)$. This expression is called *Bayes' formula*.

21. *Random variable*: *A random variable* X is a function defined on the sample space I, such that for every sample point e in I there is a determined real number (denoted as $X(e)$) corresponding to it.

22. *Mathematical expectation*: Let X be a random variable. Then the expression

$$E(X) = \sum_{e \in I} X(e) \cdot P(e)$$

is called *the mathematical expectation* of X, where e runs through every sample point in I. Let a be a constant number. We have

$$E(aX) = a \cdot E(X).$$

If X and Y are two random variables, then we have

$$E(X+Y) = E(X) + E(Y).$$

1. Coin Tossing

The first round of 2004 Europe Soccer Cup is Greece versus Portugal. At the beginning of the game, the referee let the two team captains come up and asked them to decide who should guess the result of coin tossing. Then he took out a coin and tossed it into the air. The result is the head side faced up on the ground, which is in accordance with what the Greece team captain had guessed. Therefore, the captain decided which goal his team will attack in the first half of the match. The game ended with a surprising result that Greece defeated Portugal by 2 : 1.

Assuming that the coin is made evenly, therefore, when it reaches the ground, the chance of the head side up and that of the tail side up are the same, i.e. half to half, or both is equal to $\frac{1}{2}$.

The "chance" mentioned above is called "probability" in mathematics.

Suppose there are n possible results in an experiment, and the occurrence probability of each result is the same; if among them there are m results belonging to event A, then we say the probability of A is $\frac{m}{n}$, expressed as $P(A) = \frac{m}{n}$, where $P(A)$ denotes the probability of A.

For example, when tossing a coin, there are two possible results: the head side up (denoted by A) and the tail side up (denoted by B). So $n = 2$, and

$$P(A) = P(B) = \frac{1}{2}.$$

Now tossing a coin three times, please find

(1) the probability that the head side occurs just once, and

(2) the probability that the head side occurs at least once.

Solution: (1) When tossing a coin three times, there are 8 possible results, as shown below,

HHH	HTH	THH	TTH
HHT	HTT	THT	TTT

(H=the head side up, and T=the tail side up). Among them there are 3 results belonging to the event "the head side occurs just once", so the required probability is $\frac{3}{8}$.

In general, tossing a coin n times will produce 2^n possible results, among which there are C_n^k results belonging to the event "the head side up occurs k times exactly". So the occurrence probability of this event is

$$\frac{C_n^k}{2^n}.$$

(2) Among the 8 possible results mentioned above, there is only one in which the head side does not face up even once in the three coin tosses, while in the other 7 results the head side faces up at least once. So the probability required is $\frac{7}{8}$.

In general, when tossing a coin n times, the probability that the head side does not face up even once is $\frac{1}{2^n}$, and that the head side faces up at least once is $\frac{2^n - 1}{2^n}$.

When we say that, in tossing a coin, the probabilities of the head side up and the tail side up are both equal to $\frac{1}{2}$, we do not mean that the two events will both occur $\frac{1}{2}$ time in a toss. The number of times must be a nonnegative integer. So in every time of tossing a coin, the result is either the head side up or the tail side up, and will never be half time the head side up and half time the tail side up. If tossing a coin many times, however, the number of the head side up and that of the tail side up are roughly equal. Although this conclusion seems apparent, some conscientious people still made efforts to test it. For example, the famous French scholar Georges-Louis de Buffon (1707 – 1788) had tossed a coin more than 4 thousand times, and obtained the result as shown below:

Total number of tests	The head side up	Frequency
4 040	2 048	0.506 9

Here, the frequency (of the head side up) is the ratio between the number of the head side up and that of the total tosses, i.e. $\frac{2\,048}{4\,040}$.

Another scholar, the great English statistician Karl Pearson (1857 – 1936), went even further by doing the test 2 times, and obtained the result as shown below:

Test	Total number of tosses	The head side up	Frequency
1	12 000	6 019	0.501 6
2	24 000	12 012	0.500 5

From the table above we see that the chance that the head side faces up is really about $\frac{1}{2}$, and its frequency is tending to the probability $\frac{1}{2}$ with the increase of the number of tosses.

2. General Di Qing's Coins

Di Qing (1008 - 1057), a distinguished military general of Northern Song Dynasty, was sent by the emperor to attack a powerful rebel army headed by Nungz Cigaoh (1025 - 1055) in South China. Before going out to fight the enemy, Di called together his troops and said: "Here are 100 copper coins, and I will toss them on the ground; if the result is that all the head sides face up, that means Heaven bless us and we will surely win the battle." Then he tossed the coins out. What a surprise: The 100 coins were all on the ground with the head side facing up! The soldiers burst into thunderous cheers. Di also felt very happy, and said: "Let us nail these 100 copper coins on the ground now. After defeating the enemy, we will return here to celebrate our victory."

Tossing 100 copper coins with the result that all the head sides face up on the ground, its chance is of course very small. Please find the occurrence probability of this event.

Solution: There are 2^{100} possible results in this experiment, among which only one result is all the head side facing up, so the required probability is

$$\frac{1}{2^{100}} = \left(\frac{1}{2}\right)^{100} = 7.88\ldots \times 10^{-31}.$$

By using formula $2^{10} = 1\,024 \approx 10^3$, we get a roughly estimation

$$\left(\frac{1}{2}\right)^{100} \approx 10^{-30}.$$

An event with such a small probability should happen! No wonder the soldiers thought it is the will of Heaven.

Di Qing's army won the battle as expected and come back. He ordered his soldiers to pick up the coins fixed on the ground. The soldiers found eventually that these coins are specially made: Both sides of them are the head! So it was a certain event that all of the coins, after being tossed out, lie on the ground with the head side facing up.

Even though we want just half of the 100 coins' head sides facing up, its probability is

$$\frac{C_{100}^{50}}{2^{100}} \approx 0.08,$$

which is less than $\frac{1}{10}$; therefore, it is not a common thing even for a half of the 100 coins lying on the ground with the head sides facing up.

Small probability events of course are not ones that are impossible to happen. In some important experiments, such as flying a space shuttle, we must check repeatedly to avoid hidden dangers, even though their occurrence probabilities are very small. During Space Shuttle Columbia's 28th mission on February 1, 2003, a piece of foam insulation broke off from the shuttle external tank, causing a terrible tragedy that all the 7 crews lost their lives.

A small probability event is what we say a coincidence in every day life. For example, in an NBA game, one second before the final whistle, a player seized the rebound; then, with his back to the opponent's basket, threw back the ball forcibly, and got it into the basket! It is said that the occurrence probability of this kind of event is less than a million to one. However, it really happened.

Mr. Hu Shi (1891 - 1962), a Chinese scholar with high reputation, advocated that in historical study we should "seek evidence prudently". In his writing *Research of "Dream of the Red Chamber"*, he asserted that the 40 additional chapters of the novel were written by Gao Er. His main argument is: "Cheng Wei Yuan said in his *foreword* that, he had got more than 20 additional chapters of the novel at first;

then got more than 10 other ones from a peddler. What he said itself provides firm evidence that he was lying, because there is no such a coincidence thing in the world."

The argument provided by Mr. Hu Shi is not so convincing, as it is not unusual to see a coincidence thing happened in our real life. Take Mr. Hu Shi's own example: He admitted in the appendix to his *Research of "Dream of the Red Chamber"* that he had searched for the book *Si Song Tang's Collected Poems* everywhere, with no results. "Unexpectedly, in three days I collected two versions of the book! This proves an old saying: 'You can wear out iron shoes in fruitless searching, and yet by a lucky chance you may find the lost thing without even looking for it.'"

Since Mr. Hu Shi had encountered such a coincidence thing himself, then he asserted what Cheng said is "a firm evidence that he was lying" is apparently arbitrary.

We will see in this book later that even an event with zero probability may happen in real life (e.g. in Chapter 43).

3. Rolling Dice

A dice is a small cube with a different number of spots (1 to 6) on each of its sides. We assume that the dice discussed here are all made evenly, without lead or mercury put inside. So the chances that the six possible results happen are equal when rolling a dice once, i. e. all equal to $\frac{1}{6}$.

When rolling 2 dice, there are $6 \times 6 = 36$ possible results; and rolling 3 dice, $6 \times 6 \times 6 = 216$ results. Please find the occurrence probabilities of the following events:

(1) rolling a dice and the spot number appeared is greater than 4;

(2) rolling 2 dice and the 1-spot appears at least once;

(3) rolling 3 dice and the sum of all the spot numbers appeared is greater than 15.

Solution: (1) The event that the spot number appeared is greater than 4 consists of two elementary events: the number is either 4 or 5. So the required probability is

$$\frac{2}{6} = \frac{1}{3}.$$

(2) The event that the 1-spot appears at least once consists of 11 elementary events: it appears on the first dice (6 cases) plus it appears on the second dice (6 cases) minus it appears on the both dice (1 case). So the required possibility is

$$2 \times 6 - 1 = 11.$$

(3) The event that the sum of all the spot numbers appeared is greater than 15 consists of the elementary events that the sums are

equal to 16, 17 and 18, respectively:

There is one case where the sum equals 18, i.e. the 6-spot appears on all the 3 dice.

There are 3 cases where the sums equal 17, i.e. the 6-spot appears on 2 dice and the 5-spot appears on the other one.

There are 6 cases where the sums equal 16, i.e. the spot numbers are either 6, 6 and 4, respectively (3 cases) or 6, 5 and 5, respectively (3 cases).

Therefore, there are

$$1+3+3+3=10$$

elementary events satisfying the condition. So the required probability is

$$\frac{10}{216}=\frac{5}{108}.$$

The possible results of an experiment (e.g. rolling dice) are called *sample points*, or *elementary events*. (In problem (1) above, there are 6 sample points.) The set of all the sample points is called *the sample space*. The sample spaces of problems (2) and (3) above consist of 36 and 216 sample points, respectively.

Suppose each possible result in an experiment has the same chance of appearing (in other words, the occurrence probability of each elementary event is equal). Then we say this experiment belongs to *the classical probability model*, or *the equal probability model*. Tossing coins and rolling dice mentioned above belong to the classical probability model, which is the main topic of this book, though some problems involving geometric probability will also be discussed in it.

4. Wei Xiao-bao's Bet

In the 22nd chapter of the novel *The Deer and the Cauldron* (*Lu Ding Ji*) written by Jin Yong (Louis Cha), the hero Wei Xiao-bao had a bet with Zeng Rou, a beautiful girl disciple of the Wangwu Sect, to determine the life and death of a group of 19 persons (including Zeng). The bet would be made in the following way: Wei and Zeng roll 4 dice once, respectively, to see who get the larger sum of the spot numbers (if the sum is larger than 10 or 20, then it will be subtracted 10 or 20 before comparison).

Zeng rolled dice first, and she got only 3 spots — a small number. Then it is Wei's turn, but he got 10 spots — that means zero spot, the minimum number! Zeng won the bet and the life of 19 persons were saved.

Please find the probabilities of getting a sum of 3 spots and 10 spots, respectively, in this bet.

Solution: Getting 3 spots in rolling 4 dice once means the sum of the 4 spot numbers is either 13 or 23 (it will never be 3).

(i) If the sum is 23, then the spot numbers of the 4 dice must be 6, 6, 6 and 5, respectively. Since the 5-spot can appear on any of the 4 dices, with the three 6-spots appearing on the other 3 dice, there are 4 possible results.

(ii) If the sum is 13, the distribution of the spot numbers of 4 dice will have the following 11 patterns:

(1) 6, 5, 1, 1;

(2) 6, 4, 2, 1;

(3) 6, 3, 3, 1;

(4) 6, 3, 2, 2;

(5) 5, 5, 2, 1;

(6) 5, 4, 3, 1;

(7) 5, 4, 2, 2;

(8) 5, 3, 3, 2;

(9) 4, 4, 4, 1;

(10) 4, 4, 3, 2;

(11) 4, 3, 3, 3.

Among them, (9) and (11) have 4 possible results, respectively (with the same reason in Case (i)); while(1),(3),(4),(5),(7),(8) and (10) have

$$4 \times 3 = 12$$

possible results, respectively (taking (1) as an example: firstly select a dice for the 6-spot; then select among the other three dices one for the 5-spot); while (2) and (6) have

$$4 \times 3 \times 2 \times 1 = 24$$

possible results, respectively (i. e. a full permutation of 4 different numbers).

Therefore, there are totally

$$4 + 2 \times 4 + 7 \times 12 + 2 \times 24 = 12 \times 12$$

possible results in Cases (i) and (ii). So the probability of getting a sum of 3 spots by rolling 4 dice once is

$$\frac{12 \times 12}{6^4} = \frac{1}{9}.$$

In a similar way, we can find that there are $80 + 35 = 115$ possible results in rolling 4 dice once with a sum of 10 spots (the detailed computation is left to the readers as an exercise).

However, according to the traditional Chinese rules for rolling 4 dice: Two 6-spot dice together are called *a Tianpai*, and two 1-spot dice together called *a Dipai*; if a result of rolling 4 dice has a sum of 20 spots and contains a Tianpai, then it is called *a Tiangang*; if it has a sum of 10 spots and contains a Dipai, then it is called *a Digang*; neither Tiangang nor Digang is recognized as having a 0-spot sum. Besides, if a result contains four 5-spot dice, then it is called *a pair of Changpais* (two 5-spot dice constitute *a Changpai*); and if it contains of a 1-, 2-, 3-and 4-spot dice, respectively, then it is called *a pair of Zawus* (4 and 1, as well as 3 and 2, constitute *a Zawu*); also, neither Changpai nor Zawu is recognized as having a 0-spot sum.

Consequently, there are only 5-spot number distribution patterns for getting a sum of zero spot by rolling 4 dice once, as shown below.

(1) 6, 5, 5, 4;

(2) 5, 2, 2, 1;

(3) 4, 2, 2, 2;

(4) 3, 3, 3, 1;

(5) 3, 3, 2, 2.

(According to the novel, Wei Xiao-bao has rolled out the result of pattern (2)). Then there are totally

$$2 \times 12 + 2 \times 4 + C_4^2 = 38$$

possible results with a sum of zero spot in rolling 4 dice once. So the probability of getting a sum of 3 spots by rolling 4 dice once is

$$\frac{38}{6^4} = \frac{19}{648}.$$

As a matter of fact, to get a sum of zero spot is more difficult than to get that of 3 spots in this bet, as the probability value of $\frac{144}{6^4}$ is much more than that of $\frac{38}{6^4}$.

5. Hold All the Trump Cards

In a bridge card game, 52 cards are dealt evenly to four players A, B, C, D. Please find the probability that the 13 cards that player A holds is in the same suit.

Solution: As the 13 cards player A holds are from 52 ones, there are C_{52}^{13} possible results, among which only 4 cases satisfy the condition that they are in the same suit (club, diamond, heart or spade). So the required probability is

$$\frac{4}{C_{52}^{13}} = \frac{4 \times 13! \times 39!}{52!}$$

$$= \frac{1}{50 \times 49 \times 47 \times 46 \times 43 \times 41 \times 17}$$

$$= 6.299\,078\ldots \times 10^{-12}.$$

It is a tiny number, so it is a rare event for a player to hold all the trump cards in a game.

Another way of solution is to calculate the probabilities of getting the required 13 cards one by one: The first card can be any of the 52 ones (with probability 1 or $\frac{52}{52}$); the second card, got from the remaining 51 ones, must be in the same suit as the first (with probability $\frac{12}{51}$); in a similar way, the probability that the third card is in the same suit as the two previous ones is $\frac{11}{50}$ and so on. Therefore the required probability is

$$1 \times \frac{12}{51} \times \frac{11}{50} \times \frac{10}{49} \times \frac{9}{48} \times \frac{8}{47} \times \frac{7}{46} \times \frac{6}{45} \times \frac{5}{44} \times \frac{4}{43} \times \frac{3}{42} \times \frac{2}{41} \times \frac{1}{40}$$

$$= \frac{1}{50 \times 49 \times 47 \times 46 \times 43 \times 41 \times 17}.$$

The result is the same as we got above.

The second solution has used the rule of multiplication of probabilities, i.e.

$$P(AB) = P(A) \cdot P(B \mid A),$$

where $P(A)$ is the occurrence probability of event A, $P(B \mid A)$ is the occurrence probability of event B under the condition that A happens, and $P(AB)$ is the probability the both A and B occur.

In general,

$$P(A_1A_2 \ldots A_n)$$
$$= P(A_1) \cdot P(A_2 \mid A_1) \cdot P(A_3 \mid A_1A_2) \ldots P(A_n \mid A_1A_2 \ldots A_{n-1}).$$

Taking the problem above as an example, A_1 is the event of getting the first card, A_2 is that of getting the second card in the same suit as the first, A_3 the third card in the same suit as the previous two and so on, and A_{13} the thirteen card in the same suit as the previous twelve.

It is not necessary to learn this formula by rote, as the most important thing is to understand its meaning and to be good at using it. It is essentially the multiplication principle in permutation and combination (you can have a comparison of them).

6. Roll One-Spot

When rolling a dice, how many times are needed so that the probability of getting the 1-spot at least once is greater than $\frac{1}{2}$?

Solution: Rolling a dice once, the probability of not getting the 1-spot is $\frac{5}{6}$.

Rolling a dice r times, the probability of not getting the 1-spot even once is

$$\underbrace{\frac{5}{6} \times \frac{5}{6} \times \cdots \times \frac{5}{6}}_{r} = \left(\frac{5}{6}\right)^r.$$

Therefore, rolling a dice r times, the probability of getting the 1-spot at least once is

$$1 - \left(\frac{5}{6}\right)^r. \tag{1}$$

When $r = 1, 2, 3$ and 4, the probabilities are $\frac{1}{6}$, $\frac{11}{36}$, $\frac{91}{216}$ and $\frac{671}{1\,296}$, respectively. So when $r \geqslant 4$, the value of expression (1) will be greater than $\frac{1}{2}$ (obviously the value will increase as r increases).

In a rolling dice problem like this, the events of getting and not getting the 1-spot are called *complementary events*, which means: one of the two events must occur while they cannot occur simultaneously. The sum of the probabilities of a pair of *complementary events* is 1. In other words, if A and B are complementary events, then

$$P(A) + P(B) = 1, \tag{2}$$

where $P(A)$ and $P(B)$ are the occurrence probabilities of A and B, respectively. The problem above is solved by using this formula. Whenever it is difficult to find the value of $P(A)$, we can find first the probability of A's complementary event B, and then, by using formula (2) we get

$$P(A) = 1 - P(B).$$

The complementary event of A is usually denoted as $\bar{A}$.

7. Red Balls and Black Balls

There are 4 red balls and 2 black balls in a bag, all being of the same size. Now draw one ball (not putting it back) two times. Please find the probabilities of the following events:

(1) the two balls drawn are both red;

(2) the two balls are in the same color;

(3) at least of one of the two balls is red.

Furthermore, if the first ball drawn is put back into the bag before the second draw, what are the answers to the three questions above?

Remark: This problem involves two drawing modes: the first one is called *sampling without replacement*, and the second called *sampling with replacement*.

Solution: We first solve the problem under the condition of the first drawing mode.

(1) There are $C_4^2 = \frac{4 \times 3}{2}$ ways of drawing 2 balls out of 4 red ones, and $C_6^2 = \frac{6 \times 5}{2}$ ways of drawing 2 balls out of 6 ones. So the required probability is

$$\frac{C_4^2}{C_6^2} = \frac{4 \times 3}{6 \times 5} = \frac{2}{5}.$$

Another way of solution is: The probability of the first ball drawn being red is $\frac{4}{6}$, and that of the second ball being red is $\frac{3}{5}$. Then the required probability is

$$\frac{4}{6}\times\frac{3}{5}=\frac{2}{5}.$$

(2) There are C_4^2 ways of drawing 2 balls out of 4 red ones, and C_2^2 ways of drawing 2 balls out of 2 black ones. Then there are totally

$$C_4^2+C_2^2$$

ways of drawing 2 balls with the same color out of 6 ones. Since there are C_6^2 ways of drawing 2 balls out of 6 ones, so the required probability is

$$\frac{C_4^2+C_2^2}{C_6^2}=\frac{12+2}{30}=\frac{14}{30}=\frac{7}{15}.$$

Another way of solution is: Since the probability of the two balls drawn being black is

$$\frac{C_2^2}{C_6^2}=\frac{2}{6\times 5}=\frac{1}{15}\left(\text{or } \frac{2}{6}\times\frac{1}{5}=\frac{1}{15}\right),$$

combining it with the result of (1), we get

$$\frac{12}{30}+\frac{2}{30}=\frac{14}{30}=\frac{7}{15}.$$

In the latter solution, the addition rule of probabilities is used, i.e. when A and B are *mutually inconsistent events*, we have

$$P(A\cup B)=P(A)+P(B),$$

where $P(A\cup B)$ denotes the probability that either A or B must occur. When we say that events A and B are *mutually inconsistent*, we mean that A and B cannot occur simultaneously (in our problem, for example, events A, B, and $A\cup B$ denote that the two balls are red, black and of the same color, respectively).

In general, when events A_1, A_2, ..., A_n are mutually inconsistent one another, we have

$$P(A_1\cup A_2\cup\cdots A_n)=P(A_1)+P(A_2)+\cdots+P(A_n).$$

This formula is, as a matter of fact, the additional principle in permutations and combinations.

Note that complementary events must be mutually inconsistent, while mutually inconsistent events may not be complementary. For example, the events that 2 balls drawn are red and that they are black are mutually inconsistent but not complementary, as the 2 balls drawn may be in different colors (one is red and the other black). The union of a pair of complementary events A and B must be the whole sample space (i.e. $P(A \cup B) = P(A) + P(B) = 1$), while the sum of the probabilities of mutually inconsistent events may not equal 1.

(3) The probability that 2 balls drawn are both black is $\frac{2}{30}$. So the probability that at least one of the 2 balls is red is

$$1 - \frac{2}{30} = \frac{14}{15}.$$

The details of the solution to the problem under the condition of the second drawing mode are omitted here, while the required probabilities are given below.

(1) $\frac{4}{6} \times \frac{4}{6} = \frac{16}{36} = \frac{4}{9}$;

(2) $\frac{2}{6} \times \frac{2}{6} + \frac{4}{6} \times \frac{4}{6} = \frac{20}{36} = \frac{5}{9}$;

(3) $1 - \frac{2}{6} \times \frac{2}{6} = \frac{8}{9}$.

8. Same Month and Day

One of the most famous story in Chinese classic novel *The Three Kingdoms* is that of *Peach Garden Sworn Brothers*: Three men — Liu, Guan and Zhang — swore to become brothers, by stating unanimously, "We are not seeking being born the same year, month, and day, but are only willing to die the same year, month, and day".

As a fact of matter, it is not so difficult to find persons born "the same month and day". According to the drawer principle, there are certainly two persons whose birthdays meet the condition of "the same month and day" among any 366 people born in a common year (367 people, when in a leap year).

However, if we do not demand it be a certain event, but that its probability greater than $\frac{1}{2}$; then how many people are needed to find out among them two persons born "the same month and day"?

The required number might be much smaller than you guessed: We need only 23 people to ensure that the probability is greater than $\frac{1}{2}$.

Suppose there are r people, and let n represent number 366 or 365. Then each of the r people has n possible birthdays, and there are totally n^r possible choices.

In order that their birthdays are different from each other, the first person's birthday has n choices, the second has $n-1$, the third $n-2$, and so on, and the last person has only $n-r+1$ choices; therefore, there are totally $n(n-1)(n-2)\cdots(n-r+1)$ (i.e. the permutation number P_n^r) admitted choices. So the required probability is

$$\frac{n(n-1)(n-2)\cdots(n-r+1)}{n^r}. \tag{1}$$

From (1), we have that the probability that there are two persons born 'the same month and day' among r people is

$$1-\frac{n(n-1)(n-2)\cdots(n-r+1)}{n^r}. \tag{2}$$

To let the value of expression (2) be greater than $\frac{1}{2}$, we have equivalently

$$\frac{n(n-1)(n-2)\cdots(n-r+1)}{n^r}<\frac{1}{2}. \tag{3}$$

The value of the left side of inequality (3) decreases as r increases, because $\frac{n-r+1}{n}<1$. For $n=366$, inequality (2) holds when $r=23$ (the value of the left side is 0.493 67...); for $n=365$, inequality (2) also holds when $r=23$ (the value of the left side is 0.492 70..., but it is 0.524 30..., when $r=22$).

Consequently, the probability that there are two persons born 'the same month and day' among 23 people is greater than $\frac{1}{2}$.

If we demand that the probability there are three persons born 'the same month and day' be greater than $\frac{1}{2}$, then how many people are needed to meet the demand?

In a similar way, suppose there are r people, and let n represent number 366 or 365. Then we need to exclude all the cases where there are no three persons born the same month and day.

For the case that all the people have different birthday with each other, the corresponding probability is expressed by (1).

For the case that there is exactly a pair of two persons born the same month and day among them, the corresponding probability is

$$C_r^2\times\frac{n(n-1)(n-2)\cdots(n-r+2)}{n^r}. \tag{4}$$

For the case that there are exactly two born-the-same-month-and-day pairs (but people of different pairs were born in different days), the probability is

$$\frac{1}{2!}C_r^2C_{r-2}^2\frac{n(n-1)\cdots(n-r+3)}{n^r}.$$

In general, for the case where there are exactly $k(2k \leqslant r)$ born-the-same-month- and-day pairs (but people of different pairs were born in different days), the probability is

$$\frac{1}{k!}C_r^2C_{r-2}^2\cdots C_{r-2k+2}^2\frac{n(n-1)\cdots(n-r+k+1)}{n^r}.$$

Therefore, the probability that there are three persons born "the same month and day" among any r people is

$$1-\frac{n(n-1)(n-2)\cdots(n-r+1)}{n^r}-\sum_{k=1}^{\left[\frac{r}{2}\right]}\frac{1}{k!}C_r^2C_{r-2}^2\cdots C_{r-2k+2}^2\frac{n(n-1)\cdots(n-r+k+1)}{n^r}. \quad (5)$$

By a computer algorithm, we get the following value table of expression (5).

	$n=365$	$n=366$
$r=56$	0.171 621	0.172 441
$r=100$	0.643 975	0.645 666
$r=112$	0.761 758	0.763 505
$r=150$	0.964 165	0.964 767

From the table we know that: when $r=100$, the corresponding probability is greater than $\frac{1}{2}$; when $r=150$, the event that there are three people born the same month and day will happen almost surely.

9. Integer Divisibility

Selecting randomly an integer from 1 to 2 000, please find the probability that neither 6 nor 8 divides the integer.

Solution: Let A denote the event "the integer selected is divided by 6" and B the event "the integer selected is divided by 8". Since $333 < \frac{2\,000}{6} < 334$, there are exactly 333 integers divisible by 6 between 1 and 2 000; then we have

$$P(A) = \frac{333}{2\,000}. \tag{1}$$

In a similar way, since $\frac{2\,000}{8} = 250$, there are exactly 250 integers divisible by 8 between 1 and 2 000; therefore

$$P(B) = \frac{250}{2\,000}. \tag{2}$$

As the least common multiple of 6 and 8 is 24 and $83 < \frac{2\,000}{24} < 84$, there are exactly 83 integers between 1 and 2 000 divisible by 24 (i.e. divisible both by 6 and 8); then we have

$$P(AB) = \frac{83}{2\,000}. \tag{3}$$

Therefore,

$$P(A \cup B) = P(A) + P(B) - P(AB) \tag{4}$$

$$= \frac{333}{2\,000} + \frac{250}{2\,000} - \frac{83}{2\,000} = \frac{1}{4}. \tag{5}$$

Consequently, the required probability is

$$P(\bar{A} \cap \bar{B}) = P(\overline{A \cup B}) = 1 - \frac{1}{4} = \frac{3}{4}. \quad (6)$$

Since A and B are not mutually inconsistent events, we cannot use the formula $P(A \cup B) = P(A) + P(B)$, but use a more general formula expressed in (4) instead. Formula (4) is essentially an application of the inclusion-exclusion principle in combinatorics, and it tells us that there are

$$333 + 250 - 83 = 500$$

integers divisible by either 6 or 8 between 1 and 2 000; then the probability the integer selected is divisible by either 6 or 8 is

$$P(A \cup B) = \frac{500}{2\,000} = \frac{1}{4}.$$

This example also tells us that $P(AB) = P(A) \cdot P(B)$ may not hold (unless A and B are independent with each other).

10. Repeated Experiments

Rolling 2 dice 4 times, we ask: what is the probability that the dice have the same spot number exactly twice.

Solution: Rolling 2 dice once will produce $6 \times 6 = 36$ possible results, among which there are 6 ones (i.e. two 1-spots, two 2-spots, two 3-spots, two 4-spots, two 5-spots, and two 6-spots) satisfying the condition that the dice have the same spot number; therefore, its probability is $\frac{6}{36} = \frac{1}{6}$, while that the dice have different spots is

$$1 - \frac{1}{6} = \frac{5}{6}.$$

In the process of playing 4 times, there are C_4^2 possible arrangements in which the dice have the same spot number exactly twice. For example, one of such a kind of arrangements is: the dice have the same spot number in the first and second rollings, and have different spot numbers in the third and fourth rollings. In this case, the probability that the dice have the same spot number in the first and second rolling is

$$\frac{1}{6} \times \frac{1}{6},$$

and that the dice have different spots in the third and fourth times is

$$\frac{5}{6} \times \frac{5}{6}.$$

Therefore, the probability that this arrangement occurs is

$$\left(\frac{1}{6}\right)^2 \times \left(\frac{5}{6}\right)^2.$$

In a similar way, we can show that it is also the occurrence probability for any of the other 5 arrangements. So we get finally the required probability as

$$C_4^2 \times \left(\frac{1}{6}\right)^2 \times \left(\frac{5}{6}\right)^2 = \frac{25}{216}. \tag{1}$$

In general, suppose the probability that event A occurs in an experiment is p and its nonoccurrence probability is q ($q = 1 - p$). Then in n experiments of the same kind, the probability that A occurs exactly k times is

$$C_n^k p^k q^{n-k}. \tag{2}$$

Of course we assume here that these n experiments are independent with each other (i. e. the former experiments will not affect the latter ones); otherwise, formula (2) cannot be used.

Formula (2) is a term in the expansion of the binomial expression $(p + q)^n$.

Formula (2) is frequently used in the classic probability theory. Taking the problem above as an example, if we want to find the probability that the dice have the same spot number at least twice, we need only to find the sum of the first three terms in the expansion of $\left(\frac{1}{6} + \frac{5}{6}\right)^4$; that is

$$\frac{1}{6^4} + 4 \times \left(\frac{1}{6}\right)^3 \times \left(\frac{5}{6}\right) + 6 \times \left(\frac{1}{6}\right)^2 \times \left(\frac{5}{6}\right)^2 = \frac{19}{144}.$$

11. Silver Medal Dream

The table tennis men's single competition in Olympic Games adopts elimination system (i. e. a player will be eliminated after losing once). That means the second-ranked player may be unfortunately eliminated before he meets the first-ranked one in the final round, despite the fact that, according to his power, he should win the silver medal.

Suppose there are 32 players attending the competition, and by drawing lots, the upper half competition program is scheduled as follows:

1 2 3 4 5 6 7 8 9 10 11 12 13 14 15 16

The bottom half program is similar and is omitted here. In the final round, the only surviving player from the upper half will have a match with that from the bottom half, to determine the first and second places winners of the competition. We ask: what is the probability that the second-ranked player realize his silver medal dream?

Solution: Without loss of generality, we assume the first-ranked player occupies the No. 1 place. Then the second-ranked player has the other 31 places for choice. He has to stay in the bottom half in order to realize his silver medal dream; therefore, the required probability is $\frac{16}{31}$.

In general, if there are 2^n players attending the competition, then the probability that the second-ranked player realizes his dream is

$$\frac{2^{n-1}}{2^n - 1}. \tag{1}$$

When the number m of players is not a power of 2, or in other word

$$2^{n-1} < m < 2^n, \tag{2}$$

we usually add $2^n - m$ virtual players to ensure 2^n ones. Whenever a real player meets a virtual one, he will have a bye, and enter the next round automatically. So in this situation, the probability that the second-ranked player realizes his dream remains to be (1) with n satisfying (2).

12. Fight Between Brothers

The Knights of the Round Table in King Arthur's Court is a well-known British legend. Suppose there are 2^n Knights and they will fight with each other: the losers are eliminated and the winners enter the next round of fighting. We ask: what is the probability that Balin and Balan, a pair of twin Knights, meet in the fighting? (According to the legend, the twin Knights were both killed by the other in fighting — A really bad thing! In our cases, however, the result of a fight is always one winner and one loser, no tie nor both killed. In addition, it is assumed that the win probability of each Knight is equally $\frac{1}{2}$.)

Solution: Firstly, we look into a few simple cases.

When $n = 1$, there are only 2 Knights, so the twin will certainly meet each other.

When $n = 2$, we may assume that the elder brother occupies No. 1 place (see the table of competition program in Chapter 11). Then the probability that the brothers meet each other in the first round is $\frac{1}{3}$ (the young brother occupies No. 2 place, not No. 3 or No. 4 one), and that in the second round is

$$\frac{2}{3} \times \left(\frac{1}{2}\right)^2 = \frac{1}{6}.$$

(The young brother occupies No. 3 or No. 4 place, and they have both won in the first round.)

Therefore, the required probability is

$$\frac{1}{3} + \frac{1}{6} = \frac{1}{2}.$$

When $n = 3$, in a similar way, the probability that the twins meet in fighting is

$$\frac{1}{7} + \frac{2}{7} \times \left(\frac{1}{2}\right)^2 + \frac{4}{7} \times \left(\frac{1}{2}\right)^4 = \frac{1}{4}.$$

By now, we guess that the general result is $\frac{1}{2^{n-1}}$. We are going to prove it by mathematical induction.

Assuming that it is true for the case of $n-1$, let us consider the case of n.

We still assume that the elder brother occupies No. 1 place. Then there are 2 possible choices for the young brother:

(1) He is also in the upper half. The probability of this event to occur is $\frac{2^{n-1}-1}{2^n-1}$ (occupying a place between the No. 2 and No. 2^{n-1} from No. 2 to No. 2^n ones), and by inductive assumption the probability they meet is $\frac{1}{2^{n-2}}$.

(2) He is in the bottom half. The probability of this event occurs is $\frac{2^{n-1}}{2^n-1}$, and each of them has to win $n-1$ fightings (with the probability $\left(\frac{1}{2^{n-1}}\right)^2$) before they meet in the final round. Consequently, the required probability is

$$\begin{aligned} &\frac{2^{n-1}-1}{2^n-1} \times \frac{1}{2^{n-2}} + \frac{2^{n-1}}{2^n-1} \times \left(\frac{1}{2^{n-1}}\right)^2 \\ &= \frac{1}{2^n-1} \times \frac{1}{2^{n-1}} \times (2 \times (2^{n-1}-1)+1) \\ &= \frac{1}{2^{n-1}}. \end{aligned}$$

13. Subject Groups

Three subject groups are to be formed randomly by 15 students (including 3 girls), under the condition that each group consists of 5 students and each student attends only one group. Please find the probabilities of the following events.

(1) There is exactly one girl in each group.

(2) The 3 girls attend the same group.

Solution: We may at first select 5 students to form Group 1, and select another 5 ones to form Group 2, with the remaining 5 forming the last group automatically. So there are

$$C_{15}^{5} \times C_{10}^{5} = \frac{15!}{5!5!5!} \tag{1}$$

choices to form 3 subject groups.

(1) There are 3! choices to distribute three girls to three groups with each group having exactly one girl; and there are $\frac{12!}{4!4!4!}$ choices to distribute 12 boys into three groups with each group having exactly 3 boys. So there are totally

$$3! \times \frac{12!}{4!4!4!} \tag{2}$$

choices satisfying the requirement. Therefore, the probability is

$$3! \times \frac{12!}{4!4!4!} \div \frac{15!}{5!5!5!} = \frac{25}{91}.$$

(2) There are 3 choices for three girls to attend the same group, and, once it is settled, there are

$$C_{12}^{2} \times C_{10}^{5}$$

choices for the remaining 12 boys to attend the 3 groups. So there are totally

$$3 \times C_{12}^{2} \times C_{10}^{5} = 3 \times \frac{12!}{2!5!5!}$$

choices satisfying the requirement. Therefore, the probability is

$$3 \times \frac{12!}{2!5!5!} \div \frac{15!}{5!5!5!} = \frac{6}{91}.$$

14. More Dice

Usually the number of dice used in a game will not exceed 6. One day, however, a gambler proposed on a whim: How about playing with 12, 18 or more dice?

Some interesting probability problems arises from the gambler's proposal. For example, among the following events:

roll 6 dice and get at least one 6-point,

roll 12 dice and get at least two 6-points,

roll 18 dice and get at least three 6-points, and so on,

which one gets the largest occurrence probability?

Solution: In rolling 6 dice, the probability of getting j 6-points is

$$C_6^j \times \left(\frac{1}{6}\right)^j \times \left(\frac{5}{6}\right)^{6-j},$$

then that of getting at least one 6-point is

$$\sum_{j=1}^{6} C_6^j \times \left(\frac{1}{6}\right)^j \times \left(\frac{5}{6}\right)^{6-j} = 1 - \left(\frac{5}{6}\right)^6 \approx 0.665.$$

In rolling 12 dice, the probability of getting at least two 6-points is

$$1 - \left(\frac{5}{6}\right)^{12} - 12 \times \frac{1}{6} \times \left(\frac{5}{6}\right)^{11} \approx 0.619.$$

In rolling 18 dice, the probability of getting at least three 6-points is

$$1 - \left(\frac{5}{6}\right)^{18} - 18 \times \frac{1}{6} \times \left(\frac{5}{6}\right)^{17} - \frac{18 \times 17}{2} \times \left(\frac{1}{6}\right)^2 \times \left(\frac{5}{6}\right)^{16} \approx 0.597.$$

We can see from above that, as the number of dice increases, the probability decreases. Furthermore, since it is difficult for one to hold too many dice with a hand, it is impracticable to play with more than 18 dice.

It is said that this problem was raised by a gambler to Sir Isaac Newton (1643 - 1727), the great English mathematician and physicist. As there was neither computer nor mathematical chart for use at that time, it is not easy for Newton to solve the problem.

15. Custodian Turned Thief

There was a kingdom, where misappropriation prevailed. One day, the king sent 100 boxes, each containing 100 gold ingots, to the Chancellor of the Exchequer, who was said to be one of the most honest man in the kingdom, for preservation. The chancellor then took out a gold ingot from a box and put back a false one each day. In this way, he had stolen 100 gold ingots from the 100 boxes in 100 days. Suddenly, the king came. He took out and checked an ingot from every box. We ask: What is the probability that the king did not discover the chancellor's embezzlement?

Solution: The probability of discovering the false gold ingot in a box is $\frac{1}{100}$, and that of not discovering is $1-\frac{1}{100}$. Then, after checking 100 boxes, the probability that the king did not discover the chancellor's embezzlement is

$$\left(1-\frac{1}{100}\right)^{100} = 0.36603\ldots$$

approximately $\frac{1}{3}$.

The chancellor could get so many gold ingots without any cost and with the risk of being discovered less than $\frac{2}{3}$; no wonder he would turn himself from a custodian to a thief. However, he stole only one gold ingot each time, so he was indeed "the most honest man" in the kingdom.

In general, when we replace 100 with n, the required probability will be

$$\left(1-\frac{1}{n}\right)^{n}.$$

As $n \to \infty$, we have

$$\left(1-\frac{1}{n}\right)^{n} \to \frac{1}{\mathrm{e}} = 0.367\,879\ldots$$

(where e is the base of natural logarithm, i.e. 2.718 281 828 459 045...), and that is still more than $\frac{1}{3}$. (We can prove that the probability $\left(1-\frac{1}{n}\right)^{n}$ increases with the increase of n.)

16. Put Back or Not

In a party, the host presents two pots with the same appearance, and tells you: there are two red and one black balls in one pot (named as A), and 101 red and 100 black balls in the other pot (named as B). Then he takes out a ball from one of the pots, showing it to you, and asks you whether put it back or not. After acting according to your answer, he takes out again a ball from the same pot, showing it to you, and asks you whether the pot he has drawn a ball from in the last time is A or B.

In order to get a larger probability that you guess correctly, should your first answer be to put the ball back or not?

Solution: When your answer is to put the ball back, we have the following table.

Balls taken out	The probability they are from pot A		The probability they are from pot B	You should guess
Red, red	$\frac{1}{2}\times\frac{2}{3}\times\frac{2}{3}^{*}$	$>$	$\frac{1}{2}\times\frac{101}{201}\times\frac{101}{201}$	A
Red, black	$\frac{1}{2}\times\frac{2}{3}\times\frac{1}{3}$	$<$	$\frac{1}{2}\times\frac{101}{201}\times\frac{100}{201}^{*}$	B
Black, red	$\frac{1}{2}\times\frac{1}{3}\times\frac{2}{3}$	$<$	$\frac{1}{2}\times\frac{100}{201}\times\frac{101}{201}^{*}$	B
Black, black	$\frac{1}{2}\times\frac{1}{3}\times\frac{1}{3}$	$<$	$\frac{1}{2}\times\frac{100}{201}\times\frac{100}{201}^{*}$	B

The probability that you guess correctly is the sum of the four ones attached with "$*$" in the table above. That is

$$\frac{1}{2}\times\left(\frac{2}{3}\times\frac{2}{3}+\frac{101}{201}\times\frac{100}{201}+\frac{100}{201}\times\frac{101}{201}+\frac{100}{201}\times\frac{100}{201}\right)=0.595\,97\ldots.$$

When the answer is not to put the ball back, we have the following table.

Balls taken out	The probability they are from pot A		The probability they are from pot B	You should guess
Red, red	$\frac{1}{2}\times\frac{2}{3}\times\frac{1}{2}$	$>$	$\frac{1}{2}\times\frac{101}{201}\times\frac{100}{200}$	A
Red, black	$\frac{1}{2}\times\frac{2}{3}\times\frac{1}{2}$	$>$	$\frac{1}{2}\times\frac{101}{201}\times\frac{100}{200}$	A
Black, red	$\frac{1}{2}\times\frac{1}{3}\times 1$	$>$	$\frac{1}{2}\times\frac{100}{201}\times\frac{101}{200}$	A
Black, black	0	$<$	$\frac{1}{2}\times\frac{100}{201}\times\frac{99}{200}$	B

The probability that you guess correctly is

$$\frac{1}{2}\times\left(\frac{1}{3}+\frac{1}{3}+\frac{1}{3}+\frac{100}{201}\times\frac{99}{200}\right)=0.623\,13\ldots.$$

Therefore, the probability that you guess correctly when you let the ball back is larger than that when you do not let the ball back.

Another way of answer better than the previous two is: If the first ball is red, your answer is to put it back; otherwise, your answer is not to put it back. In this case, we have the following table.

Balls taken out	The probability they are from pot A		The probability they are from pot B	You should guess
Red, red	$\frac{1}{2}\times\frac{2}{3}\times\frac{2}{3}$	$>$	$\frac{1}{2}\times\frac{101}{201}\times\frac{101}{200}$	A
Red, black	$\frac{1}{2}\times\frac{2}{3}\times\frac{1}{3}$	$<$	$\frac{1}{2}\times\frac{101}{201}\times\frac{100}{201}$	B
Black, red	$\frac{1}{2}\times\frac{1}{3}\times 1$	$>$	$\frac{1}{2}\times\frac{100}{201}\times\frac{101}{200}$	A
Black, black	0	$<$	$\frac{1}{2}\times\frac{100}{201}\times\frac{99}{200}$	B

The probability that you guess correctly is

$$\frac{1}{2}\times\left(\frac{4}{9}+\frac{1}{3}+\frac{101}{201}\times\frac{100}{201}+\frac{100}{201}\times\frac{99}{200}\right)=0.63702\ldots.$$

It is larger than the probabilities you get in previous two cases.

It is suggested that our readers calculate the probability of guessing correctly in the case: if the first ball is red, your answer is not to put it back; otherwise, your answer is to put it back.

17. Match Problem

A professor has written n letters and n envelopes. His grandson puts the letters into the envelopes randomly (with each envelope containing one letter). Please find the probability that no letter matches its envelope, and the probability that there are exactly r $(1 \leqslant r \leqslant n)$ matched pairs of letters and envelopes.

Solution: There are $n!$ ways to put n letters into n envelopes separately.

There are $(n-1)!$ ways to put a specific letter into its right envelope (the other $n-1$ letters being put randomly).

There are $(n-2)!$ ways to put the second specific letter into its right envelope, and so on.

There is only one way to put all the n letters into their right envelopes separately.

By using the inclusion-exclusion principle, we get that there are

$$n! - C_n^1 \times (n-1)! + C_n^2 \times (n-2)! - \cdots + (-1)^k C_n^k \times (n-k)! + \cdots + (-1)^n \times 1$$

$$= \frac{n!}{2!} - \frac{n!}{3!} + \cdots + (-1)^k \cdot \frac{n!}{k!} + \cdots + (-1)^n \cdot \frac{n!}{n!}$$

ways such that no letter is put into its right envelope. Therefore, the probability that no letter matches its envelope is

$$\frac{1}{2!} - \frac{1}{3!} + \cdots + \frac{(-1)^k}{k!} + \cdots + \frac{(-1)^n}{n!}. \tag{1}$$

As n becomes very large, we have (1) $\to \frac{1}{\mathrm{e}} = \frac{1}{2.71828\ldots} = 0.367879\ldots$.

$\left(\text{Here } \frac{1}{e} = 1 - \frac{1}{1!} + \frac{1}{2!} - \frac{1}{3!} + \cdots + \frac{(-1)^k}{k!} + \cdots\right)$

In order to have exactly r letters put into their right envelopes, we can first select r letters from n ones, by C_n^r ways; then put the first selected letter into its right envelope, whose probability is $\frac{1}{n}$; put the second letter into its right envelope, whose probability is $\frac{1}{n-1}$ and so on; finally put the rth letter into its right envelope, whose probability is $\frac{1}{n-r+1}$. Therefore, the probability that there are exactly r matched pairs of letters and envelopes is

$$P(r, n) = C_n^r \times \frac{1}{n(n-1)\cdots(n-r+1)} \times P(0, n-r), \tag{2}$$

where $P(0, n-r)$ denotes the probability that no letter of $n-r$ ones is put into its right envelope. Thus replacing n with $n-r$ in (1), we have

$$P(0, n-r) = \frac{1}{2!} - \frac{1}{3!} + \cdots + \frac{(-1)^k}{k!} + \cdots + \frac{(-1)^{n-r}}{(n-r)!}.$$

Substituting it into (2), we have

$$\begin{aligned} P(r, n) &= \frac{1}{r!} P(0, n-r) \\ &= \frac{1}{r!}\left(\frac{1}{2!} - \frac{1}{3!} + \cdots + \frac{(-1)^k}{k!} + \cdots + \frac{(-1)^{n-r}}{(n-r)!}\right). \end{aligned} \tag{3}$$

18. Put Balls into Drawers

There are n drawers, numbered 1, 2, ..., n. Now put r ($\leqslant n$) balls into these drawers. Please find the probability that each of the first r drawers (i.e. those numbered from 1 to r) contains exactly one ball.

Solution: As these drawers are numbered, they are different from each other. But we have not been told whether the balls are different from each other, and how many balls are allowed to put into a drawer. So there are several answers according to different conditions.

When the balls are considered different from each other (e.g. they are numbered from 1 to r) and each drawer is allowed to contain any number of balls, there are n ways to put a specific ball into a drawer and then n^r ways to put all the r balls. Among them there are $r!$ ways to put exactly one ball into each of the first r drawers. Therefore, the required probability is

$$\frac{r!}{n^r}. \tag{1}$$

When the balls are considered the same and each drawer is allowed to contain at most one ball, there are C_n^r ways to distribute r balls into n drawers. And there is just one way to put one ball into each of the first r drawers. Therefore, the required probability is

$$\frac{1}{C_n^r} = \frac{r!(n-r)!}{n!}. \tag{2}$$

When the balls are considered the same and each drawer is allowed to contain any number of balls, there are C_{n+r-1}^r ways to distribute r balls into n drawers. And there is just one way to put one ball into each of the first r drawers. Therefore, the required probability is

$$\frac{1}{C_{n+r-1}^{r}}=\frac{r!(n-1)!}{(n+r-1)!}. \tag{3}$$

The problems mentioned above have been playing important roles in modern statistical physics. The three kinds of results represent three well-known theories of the science, respectively: Maxwell-Boltzmann theory, Bose-Einstein theory, and Fermi-Dirac theory.

By the way, the statistical model of putting balls into drawers, which is also called the pot problem, contains some well-known statistical problems, such as the birthday problem (where r persons may be regarded as r balls, and n day in a year as n drawers), the rolling dice problems (where rolling dice r times may be regarded as r balls, and the six number of spots from 1 to 6 as six drawers).

Note: From the $n-r+1$ places in a row, we select r places to put balls, each of them being denoted by ○, while the other places are denoted by vertical bars. We then have the following pattern:

$$|\circ|\circ\circ||\circ\circ\circ\circ\circ|||\circ|\circ \tag{4}$$

We regard the places between any two adjoining bars, as well as that before the first bar and that beyond the last one, as a drawer, respectively. Then there are exactly n drawers, each of them containing some balls or nothing. In doing so, every pattern like (4) represent a way of putting balls into drawers, and vice versa. So the number of patterns like (4), which is C_{n+r-1}^{r}, is exactly the number of ways to put r balls of the same size into n drawers.

19. Problem of Matches

A smoking mathematician has one box containing n matches in each of his left and right pockets. Every time he needs a match he will take out randomly a match box from one of the two pockets. Please find:

(i) the probability that he takes out an empty box at the first time while the other box contains exactly r ($r = 0, 1, \ldots, n$) matches;

(ii) the probability that when he takes out the last match from a box, there are exactly r matches in the other one.

Remark: These problems were proposed by the Polish mathematician S. Banach (1892 - 1945).

Solution: (i) Firstly, we assume that the mathematician takes out an empty box from his left pocket. Before it he has used totally $n + (n - r) = 2n - r$ matches, whose occurrence probability, according to the discussion in Chapter 10, is

$$C_{2n-r}^{n}\left(\frac{1}{2}\right)^{n}\left(\frac{1}{2}\right)^{n-r}.$$

At the $(2n - r + 1)$ th time he needs a match, he takes out the box in his left pocket (to realize that it is empty), with the probability $\frac{1}{2}$. So the total probability is

$$C_{2n-r}^{n}\left(\frac{1}{2}\right)^{n}\left(\frac{1}{2}\right)^{n-r}\times\frac{1}{2}. \tag{1}$$

In a similar way, the corresponding probability in the case that he takes out an empty box from his right pocket is also (1).

Consequently, the probability that he takes out an empty box at the first time while the other one contains exactly r ($r = 0, 1, \ldots, n$)

matches is

$$2 \times C_{2n-r}^{n} \left(\frac{1}{2}\right)^{n} \left(\frac{1}{2}\right)^{n-r} \times \frac{1}{2} = \frac{C_{2n-r}^{n}}{2^{2n-r}}. \tag{2}$$

Remark: When he takes out an empty box, there are $(n+1)$ possible results according to the number r ($r = 0, 1, \ldots, n$) of the matches contained in the other box. As one and only one of these mutually exclusive results must occur, we have

$$\sum_{r=0}^{n} \frac{C_{2n-r}^{n}}{2^{2n-r}} = 1 \tag{3}$$

(The probability of the certain event equals 1.) An interesting identical equation (3) is then derived by arguments from the probability theory. This identity is not so easy to prove directly.

(ii) In a similar way as shown in (i), we get that the probability that the mathematician has taken $n-1$ matches from his left (right) pocket and $n-r$ ones from his right (left) pocket before he takes the last one also from his left (right) pocket is

$$2 \times C_{2n-r-1}^{n-1} \left(\frac{1}{2}\right)^{n-1} \left(\frac{1}{2}\right)^{n-r} \times \frac{1}{2} = \frac{C_{2n-r-1}^{n}}{2^{2n-r-1}}. \tag{4}$$

And also we obtain an identical equation

$$\sum_{r=1}^{n} \frac{C_{2n-r-1}^{n}}{2^{2n-r-1}} = 1. \tag{5}$$

Usually one will throw away the match box as soon as it is empty. If so, the event that he takes out an empty box as shown in problem (i) will never happen. The event in (i) happens possibly because this mathematician like to collect match boxes.

20. Trial in a Three-Judge Court

One scene in the Peking opera *Yu Tang Chun* (*The Story of Su San*) is called *Trial in a Three-Judge Court*, where three judges — the Governor Wang Jin-long, the Ponchassi Pan Bi-zheng, and the Anchasi Liu Bing-yi — interrogated jointly Su San, an unjustly accused young woman. Wang intended to absolve Su, but Liu placed obstacles in the way. They could hardly reach an agreement on judgement.

In a modern court, sometimes there are also several judges.

Suppose there are two good judges and one bad judge in a court: a good judge will make correct judgement with the probability $p\left(\geqslant \frac{1}{2}\right)$, while a bad one, being ignorant of the law, will make a judgement according to the color of the ball he takes out from his pocket (in which 2 balls are contained: a black and a white) — therefore he will make correct judgement with the probability $\frac{1}{2}$. A trial decision is formed according to the majority opinion in the three judges.

The question is: Between the probability that the three judges jointly make the correct trial decision and that a good judge makes it alone, which is larger?

Solution: There are two cases that the three judges jointly make the correct trial decision:

(1) The two good judges both make correct judgement. The occurrence probability of this case is

$$p \times p.$$

(2) One good judge makes correct judgement and the other makes wrong one, while the bad judge makes correct judgement by the color of a ball. The probability of this case is

$$p \times (1-p) \times \frac{1}{2} + (1-p) \times p \times \frac{1}{2} = p \times (1-p).$$

Therefore, the probability that the three judges jointly make the correct trial decision is

$$p \times p + p \times (1-p) = p,$$

which is the same as that of a good judge who makes a correct one alone.

If we let two good judges jointly make a trial decision, and, when they have different opinions, let it be decided by tossing coin or taking ball, then the probability of making correct trial decision is

$$p \times p + \frac{1}{2} \times 2p(1-p) = p,$$

which is also the same as that of a good judge who makes a correct one alone.

If we let three good judges jointly make a trial decision, and, when they have different opinions, let it be decided by the majority one, then the probability of making the correct trial decision is

$$p \times p \times p + 3 \times p \times p \times (1-p) = p^2(3-2p).$$

Since $p \geqslant \frac{1}{2}$, we have

$$p(3-2p) \geqslant \frac{1}{2}\left(3 - 2 \times \frac{1}{2}\right) = 1.$$

Therefore,

$$p^2(3-2p) \geqslant p.$$

It means that three good judges jointly make a trial decision will be better than a good judge who does it alone.

21. Win Twice in Succession

All of the three people in a family can play chess. While it is not unusual for any of them to beat the other two in competitions, the father, however, is the best player among the three. One day, the son asked the father to buy a set of *Olympic Math Tutorials* for him. The latter said: "you must play chess with your mother and me in turn for three rounds; if you can win two rounds in succession, then I will buy you the books." The son then asked: "Should I first contest with you or mother?" "You can choose it", the father replied.

The question is: whom should the son choose to play chess with first?

Solution: If the son chooses to play chess with the father first, the order of his opponents is: the father, the mother, and the father. If his choice is the mother first, then the opponent order is: the mother, the father, and the mother.

By the first choice the son has to contest with the father twice, while by the second choice he needs to do so only once. Since the father is the stronger opponent, it seems the second choice is better. However, a careful calculation will show that the first one is better.

Suppose the probability the son beats the father is p, and that he beats the mother is q, where $p \leqslant q$. In the first choice, there are two cases where the son wins twice in succession:

(1) he wins the first and second rounds, with the probability $p \times q$;

(2) he loses the first round, but wins the second and third rounds, with the probability $(1-p) \times p \times q$.

So the probability for him to win twice in succession by the first choice is

$$p \times q + (1 - p) \times p \times q = pq(2 - p). \quad (1)$$

In a similar way, the required probability in the second choice is (obtained from (1) by just exchanging the places of p and q)

$$pq(2 - q). \quad (2)$$

Since $p \leqslant q$, so $2 - p \geqslant 2 - q$; therefore, the first choice is better than the second.

Why, then, the probability found in the choice of contesting with the father (the stronger player) twice should be greater than that of contesting with him only once?

The explanation lies in the fact that the probability required is about winning *two rounds in succession*: Although by the second choice the son only need to contest with his father once; however, if he is defeated in this round, he will lose the chance to win twice in succession. On the other hand, in the first choice, even if he is defeated in the first round, he still has the chance to win the second and third rounds in succession.

If the question is about the comparison of the probabilities that the son wins at least twice in these two choices, the answer will be different.

As a fact of matter, by the first choice, the son can either win twice in succession or lose the second round but win the first and third rounds. So the probability he wins at least twice is

$$pq(2 - p) + p^2(1 - q).$$

In a similar way, the probability he wins at least twice by the second choice is

$$pq(2 - q) + q^2(1 - p).$$

We have

$$\begin{aligned} & pq(2 - q) + q^2(1 - p) - pq(2 - p) - p^2(1 - q) \\ = & 2pq(p - q) + q^2 - p^2 = (q - p)(p + q - 2pq) \end{aligned}$$

$$= pq(q-p)\left(\frac{1}{q}+\frac{1}{p}-2\right).$$

Notice that $q \geqslant p$, $q \leqslant 1$ and $p \leqslant 1$, so $\frac{1}{q}+\frac{1}{p} \geqslant 2$. Therefore, the expression above is greater than zero. It means that the required probability in the second choice is larger than that in the first choice.

As for the probability that the son wins exactly twice (please note the difference in meaning between "winning exactly twice" and "winning at least twice"), it is larger also in the second choice than that in the first one. You, the reader, can prove it.

22. Fire Blank Shots

In a shooting competition at the Athens 2004 Summer Olympic Games, the American shooter Matthew Emmons was very close to winning the championship, but he accidentally cross-fired his last shot to the other player's target and finished eighth in regret.

It is naturally a very bad thing that you fire blank shots in a life-or-death duel. However, it may produce surprising results sometimes.

Let us say that Guo Jing, Zhou Bo-tong, and Qiu Qian-ren, the three characters in the novel *The Legend of the Condor Heroes* by Jin Yong, had been fighting with each other at night, with no winner and no loser. The next morning, the three men decided to duel with each other by pistol. The rule is: each man, in the order of Guo, Zhou, and Qiu, fires a shot in turn (he could choose any target); then repeat the shooting order, till there is only one man not shot (a shot man would fall down at once and of course could not shoot any more).

Suppose the probability for Guo to hit the target is 0.3, that for Zhou is 1 (he always hits the target), and that for Qiu is 0.5. The question is: which target Guo should aim at when he fires his first shot?

If Guo aims at Qiu first, the probability of hitting him is 0.3. But if Qiu is shot, Zhou will hit Guo down to the ground at once. So Qiu to not be shot is better than if he is shot. If Guo fails to hit Qiu, Zhou should shoot the more dangerous Qiu first (as Qiu shoots more accurately than Guo does); and after Qiu is shot by Zhou, Guo will have a chance of 0.3 to win (i. e. he will be the only person to survive). So the winning probability for Guo when he choose to aim at Qiu first is

$$(1-0.3)\times 0.3 = 0.7\times 0.3 = 0.21 < 0.3.$$

If Guo aims at Zhou first and hits him with the chance of 0.3, then Guo and Qiu will shoot each other in turn — the winning probability of Guo is

$$\begin{aligned}&0.5\times 0.3+0.5^2\times 0.7\times 0.3+0.5^3\times 0.7^2\times 0.3+\cdots\\&=0.5\times 0.3\times[1+0.5\times 0.7+(0.5\times 0.7)^2+\cdots]\\&=0.5\times 0.3\times\frac{1}{1-0.5\times 0.7}\\&=\frac{15}{65}=\frac{3}{13}.\end{aligned}$$

If Zhou has not been hit, then he will shoot Qiu first, and then Guo and Zhou will shoot each other with the winning chance for Guo being 0.3.

So the winning probability for Guo when he choose to aim at Zhou first is

$$0.3\times\frac{3}{13}+(1-0.3)\times 0.3. \qquad (1)$$

Since $\frac{3}{13}<0.3$, we have $(1)<0.3$.

The winning probabilities of the both choices (aim at Qiu first and aim at Zhou first) for Guo are less than 0.3.

Actually, Guo has the third choice: aim at first neither Qiu nor Zhou, but fire a blank shot. Then Zhou shoots Qiu, and he shoots Zhou in turn with a winning probability of 0.3.

Therefore, the best choice for Guo is to fire a blank shot first.

23. Catch a Turtle in a Jar

There are five jars, among which:

(1) the first and second jars each contains 2 male turtles and 1 female turtle,

(2) the third and fourth jars each contains 3 male turtles and 1 female turtle, and

(3) the fifth jar contains 10 female turtles.

Now, randomly choose a jar and randomly catch a turtle from it. Please find the probability of catching a male turtle.

Solution: The five jars are classified into three types according to the chances of catching a male turtle from them. The required chance for the first type jars is $\frac{2}{3}$; then the probability of choosing a jar of this type and catching a male turtle from it is

$$\frac{2}{5} \times \frac{2}{3}.$$

The required chances for the second and third types are $\frac{3}{4}$ and $\frac{0}{10}$, respectively; then the probabilities of choosing a jar of these types and catching a male turtle from it are

$$\frac{2}{5} \times \frac{3}{4} \text{ and } \frac{1}{5} \times \frac{0}{10},$$

respectively.

Therefore, the probability of catching a male turtle from a jar is

$$\frac{2}{5} \times \frac{2}{3} + \frac{2}{5} \times \frac{3}{4} + \frac{1}{5} \times \frac{0}{10} = \frac{17}{30}.$$

In a similar way, the probability of catching a female turtle from a jar is

$$\frac{2}{5}\times\frac{1}{3}+\frac{2}{5}\times\frac{1}{4}+\frac{1}{5}\times\frac{10}{10}=\frac{13}{30}\left(\text{i. e. } 1-\frac{17}{30}\right).$$

More generally, suppose the sample space I can be divided into a set of pairwise disjoint events $B_1, B_2, \ldots, B_n$, i.e.

$$B_1 \cup B_2 \cup \cdots \cup B_n = I, \tag{1}$$

$$B_i \cap B_j = \phi \ (1 \leqslant i < j \leqslant n). \tag{2}$$

Then the probability that event A happens is

$$P(A) = \sum_{i=1}^{n} P(A \mid B_i) \cdot P(B_i). \tag{3}$$

Expression (1) is called *the total probability formula*. We need not memorize this formula by heart. The important thing is to understand its meaning well and be good at applying it to solve problems. As a matter of fact, it is nothing more than the Rule of Product and the Rule of Sum in Permutation and Combination. It is not hard for you to solve the catching turtle problem above even though you do not know this formula.

Proof of the total probability formula: From (1) we have

$$AB_1 \cup AB_2 \cup \cdots \cup AB_n = AI = A.$$

By (2), we get $AB_i \cap AB_j \subset B_i \cap B_j = \phi$; therefore AB_1, $AB_2, \ldots, AB_n$ are pairwise disjoint. Then we have

$$P(A) = P(AB_1 \cup AB_2 \cup \cdots \cup AB_n) = \sum_{i=1}^{n} P(AB_i)$$

$$= \sum_{i=1}^{n} P(A \mid B_i) \cdot P(B_i).$$

24. Diagnosis Rate

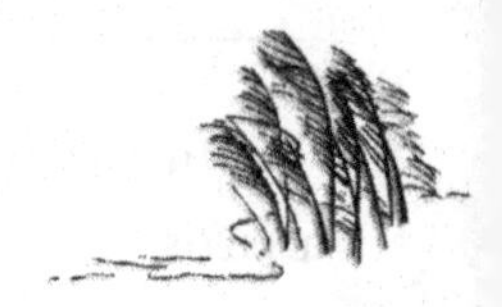

Let A denote the event of "test positive" in a specific experiment, and B that of "have cancer". It is known that $P(B) = 0.005$, $P(A \mid B) = 0.95$, and $P(\bar{A} \mid \bar{B}) = 0.95$. Suppose someone tests positive in the experiment. Please find the probability that he has cancer.

Remark: This kind of problems was first put forward by the Swiss mathematician Daniel Bernoulli (1700 - 1782), and was solved by the English statistician and minister Thomas Bayes (1701 - 1761), who proposed the famous *Bayes Formula* for doing so.

Solution: First of all, from $P(\bar{A} \mid \bar{B}) = 0.95$ we have

$$P(A \mid \bar{B}) = 1 - 0.95 = 0.05.$$

And from $P(B) = 0.005$ we get

$$P(\bar{B}) = 1 - P(B) = 0.995.$$

Therefore,

$$P(A) = P(B) \times P(A \mid B) + P(\bar{B}) \times P(A \mid \bar{B})$$

and

$$\begin{aligned} P(B \mid A) &= \frac{P(B) \times P(A \mid B)}{P(A)} \\ &= \frac{0.005 \times 0.95}{0.005 \times 0.95 + 0.995 \times 0.05} \\ &\approx 0.087. \end{aligned}$$

This means that the diagnosis rate is about 8.7% for the person who has tested positive.

From the result above we see that: Although both $P(A \mid B)$ and

$P(\bar{A} \mid \bar{B})$ are as large as 95% (that means 95% of the people who have cancer will test positive in the experiment and 95% of the people who do not have cancer will test negative), there are only 8.7% of the people who test positive will have cancer (i.e. $P(A \mid B)$ is not so large). So the people who have tested positive does not need to be very worried, presuming to have cancer themselves. On the other hand, one does not need to be very worried, even if he has cancer: He can still live very well and do many good things.

The following formula

$$P(B_1 \mid A) = \frac{P(B_1) \times P(A \mid B_1)}{\sum_{i=1}^{n} P(A \mid B_i) \cdot P(B_i)}$$

is called *the Bayes formula*, in which $B_1, B_2, \ldots, B_n$ satisfy conditions (1) and (2) in the last chapter.

The role of this formula is to find $P(B_1 \mid A)$ under the condition that $P(B_i)$ and $P(A \mid B_i)$ ($i = 1, 2, \ldots, n$) are known. It is also called *the inverse probability formula*.

The derivation of the Bayes formula is also easy: From the multiplication rule of probability, we have

$$P(B_1 \mid A) = \frac{P(AB_1)}{P(A)} = \frac{P(B_1)P(A \mid B_1)}{P(A)}.$$

Next substitute $P(A)$ by the total probability formula mentioned in the last section. The required result is then obtained.

When $n = 2$, the Bayes formula can be written as

$$P(B \mid A) = \frac{P(B) \times P(A \mid B)}{P(B) \times P(A \mid B) + P(\bar{B}) \times P(A \mid \bar{B})}.$$

25. Running Well

It is known that a particular machine will make products with a qualified rate of 90% when it is running well, but will do so with a qualified rate of only 30% when it is not running well. The probability that the machine is running well is 75% normally. Suppose that one day, the first product made by the machine is qualified. Please find the probability that the machine is running well at this time.

Remark: The running well rate 75% of the machine is obtained from the past data, and is called *a priori probability*. Now, with the condition that the first product is qualified, the probability that the machine is running well increases (on the contrary, if the first product is unqualified, the corresponding probability will decrease).

Solution: The calculation detail is presented as the following: Let A denote the event "a product is qualified", and B that "the machine is running well". It is known that $P(A \mid B) = 0.9$, $P(A \mid \bar{B}) = 0.3$, and $P(B) = 0.75$. So

$$P(\bar{B}) = 1 - P(B) = 0.25.$$

According to the Bayes formula, we have

$$\begin{aligned} P(B \mid A) &= \frac{P(A \mid B) \times P(B)}{P(A \mid B) \times P(B) + P(A \mid \bar{B}) \times P(\bar{B})} \\ &= \frac{0.9 \times 0.75}{0.9 \times 0.75 + 0.3 \times 0.25} \\ &= 0.9 = 90\%. \end{aligned}$$

This means the probability that the machine is running well is 90%.

The probability 90% is obtained after we have the information that "the first product is qualified", and is then called *a posterior* probability, which can help us to know the current state of the machine.

In a similar way, if the first product is unqualified, letting C denote the event "a product is unqualified", we then have

$$P(C \mid B) = 1 - 0.9 = 0.1$$

and

$$P(C \mid \bar{B}) = 1 - 0.3 = 0.7.$$

Therefore,

$$\begin{aligned} P(B \mid C) &= \frac{P(C \mid B) \times P(B)}{P(C \mid B) \times P(B) + P(C \mid \bar{B}) \times P(\bar{B})} \\ &= \frac{0.1 \times 0.75}{0.1 \times 0.75 + 0.7 \times 0.25} \\ &= 0.3 = 30\%, \end{aligned}$$

meaning the probability that the machine is running well is only 30%. In this situation, one should consider having an examination of the machine; otherwise, more unqualified products would be made, and the machine might be destroyed as well.

The Bayes formula has wide applications. On the other hand, however, it should not be misused. There have been frequently many disputations about how to decide a prior probability needed by the formula.

26. Money Change Problem

At the ticket window of a theater stand a queue of $2n$ people, each being allowed to buy one ticket with 5 yuan. Among them, n people each has only a 10-yuan note, and the other n people each has only a 5-yuan note. At the beginning, the ticket seller has no small money for change. Please find the probability that every person in the queue gets a ticket successfully, without having to wait because of the seller's unable to give him change temporarily.

Solution: We use a fold line in the Cartesian coordinate system to represent the state of ticket selling.

The start point of the line is the origin of the coordinate system. If the first person buys a ticket with a 5-yuan note, we connect the origin with point $(1, 1)$ in the line; if he does so with a 10-yuan note, the origin is then connected with point $(1, -1)$.

By induction, assume that the kth ($k = 1, 2, \ldots, 2n - 1$) point A_k corresponding to the buying state of the kth person is (k, y_k) in the line. The $(k + 1)$ th point A_{k+1} to be connected to A_k is determined according to the buying state of the $(k + 1)$ th person: If he uses a 5-yuan note, then $y_{k+1} = y_k + 1$; if he uses a 10-yuan note, then $y_{k+1} = y_k - 1$.

The end point of the line is $A_{2n} = (2n, y_{2n})$. It is not very difficult to see that $y_{2n} = 0$, because there are n people using a 5-yuan note and equally n people using a 10-yuan note.

The segment between A_k and A_{k+1} is represented by A_kA_{k+1} ($k = 0, 1, \ldots, 2n - 1$), and there are totally $2n$ such segments in the line. Among these segments, n ones are upward (as there are n people buying ticket with a 5-yuan note), and the other n ones are downward (there are also n people buying ticket with a 10-yuan note).

Therefore, there are C_{2n}^{n} fold lines in all, corresponding to the different distributions of the n upward segments in the $2n$ places.

These fold lines are divided into two categories: the favorite lines that lie entirely above the x-axis (meaning the seller can always give change), and the unfavorite lines that has some part lying under the x-axis (meaning the seller is unable to give change sometimes).

We now calculate the number of the unfavorite lines. Suppose the fold line fist meets straight line $y=-1$ at point A_h. Then we make a symmetry transformation about $y=-1$ of the part between A_h and the end point A_{2n} of the line, and get a new one illustrated by dots in the graph below.

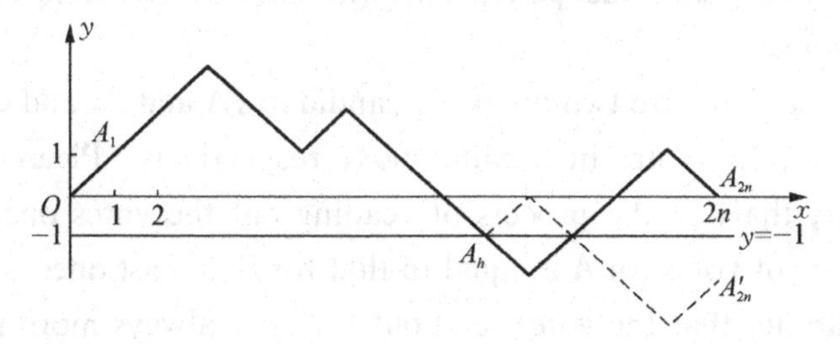

Thus, referring to the graph, we have a new fold line, whose start point is still the origin O, the part between O and A_h remains the same as the original line, and that between A_h and A'_{2n} is represented by dotted line. The end points A'_{2n} and A_{2n} of the two lines are symmetric about $y=-1$, so $A'_{2n}=(2n, -2)$. Therefore, there are $n+1$ downward segment and $n-1$ upward ones in the new fold line.

Consequently, there are C_{2n}^{n+1} unfavorite lines in all.

The number of favorite lines is then

$$C_{2n}^{n}-C_{2n}^{n+1}=C_{2n}^{n}-\frac{n}{n+1}C_{2n}^{n}=\frac{1}{n+1}C_{2n}^{n}. \tag{1}$$

Finally, the required probability is

$$\frac{\frac{1}{n+1}C_{2n}^{n}}{C_{2n}^{n}}=\frac{1}{n+1}.$$

Numbers $\frac{n}{n+1}C_{2n}^{n}$ in (1) are called the Catalan numbers, which play important roles in Combinatorial Mathematics.

27. Donkey versus Elephant

Donkey and elephant are the symbols of two major political parties in the U. S. A., the Democratic and the Republican, respectively. In every election year, the politicians will have fierce competitions in seeking votes.

Suppose there are two opposing candidates A and B, and they have a and b ($a > b$) votes in a ballot box, respectively. Please find the probability that, in the process of reading out the votes one by one, the number of votes for A is equal to that for B at least once, as well as the probability that the votes read out for A is always more than that for B.

Solution: We look into the problem in three cases.

Case 1: The first vote read out is for B (with probability $\frac{b}{a+b}$). Since $a > b$ the number of votes for A must be equal to that for B at least once in the reading out process.

Case 2: The first vote read out is for A (with probability $\frac{a}{a+b}$), and the number of votes for A equals that for B at least once in the reading out process.

Case 3: The first vote read out is for A, and the votes for A are always more than that for B in the reading out process.

Now we establish a map f from Case 2 into Case 1: For any process in Case 2, we change every vote read out before or at the moment the first equality appears to that for the opposite candidate (i.e. change a vote for A to that for B, and vice versa), and keep unchanged for the votes read out afterward. Then we get correspondingly a process in Case 1.

In this way, we can also transform any process in Case 1 to that in Case 2, and then get a map between them, which is obviously the inverse map of f.

From the discussion above, we know the number of possible reading out process in Case 2 is equal to that in Case 1. That means the probability that a process in Case 2 happens is also $\frac{b}{a+b}$.

Consequently, the probability that the number of votes for A is equal to that for B at least once in the read out process is

$$\frac{b}{a+b}+\frac{b}{a+b}=\frac{2b}{a+b}.$$

In the case that $a=b$, the equality of the votes read out for two candidates will certainly happen (at least once, e.g. at the end of the reading out process), and that means the required probability is 1. At the same time $\frac{2b}{a+b}$ also equals 1, so the formula still holds in this case.

The probability that the votes for A are always more than that for B is then

$$1-\frac{2b}{a+b}=\frac{a-b}{a+b}.$$

Remark: The method used in solving the problem here is actually very close to that in the last chapter. We can also draw a fold line as illustrated there, and at this time make a symmetry transformation of the part before the first equality about x-axis (in the last chapter, a symmetry transformation about $y=-1$ of the part after the cross point on the line $y=-1$ was made).

Of course, we can also make a symmetry transformation of the part after the first equality about x-axis (it will be further close to the case in the last chapter). At this time, the total number of possible fold lines is C_{a+b}^{a} (in each of them there are a upward segments).

Consider a favorite fold line corresponding to a reading out process in Case 2: Its first segment must be upward. Suppose it has t downward segments before the first equality. Then it will have $a - t$ downward segments afterward, as the symmetric transformation has changed the $a - t$ upward segments in the part after the first equality to downward ones. So there are totally $t + (a - t) = a$ downward segments, which can be distributed arbitrarily among the $a + b - 1$ places (except the first place) in the line. Therefore, there are C_{a+b-1}^{a} possible favorite lines, and that means the probability that a process in Case 2 occurs is

$$\frac{C_{a+b-1}^{a}}{C_{a+b}^{a}} = \frac{b}{a+b}.$$

28. East Wind versus West Wind

Lin Dai-yu, a girl heroine in the famous Chinese classic novel *A Dream of Red Mansions*, once said: Either East Wind overwhelms West Wind or West Wind overwhelms East Wind.

Suppose East Wind contests with West Wind n times, the numbers of times they win over each other in the process of competitions are recorded, and in each competition they have an equal chance of 50% to win respectively. If the number of times East Wind wins is always greater than that West Wind does, we then say that East Wind overwhelms West Wind.

Please find the probability that East Wind overwhelms West Wind in this series of competitions.

Solution: If East Wind always wins more than West Wind does in the process of competitions, it will never appear that the numbers of winnings recorded for them respectively are equal.

On the other hand, if the recorded winning numbers of them are not equal even once in the process of competitions, it must be as what Lin Dai-yu said, "Either East Wind overwhelms West Wind or West Wind overwhelms East Wind". In this case, the occurrence probabilities of the two events must be a half each, as they are equally matched opponents.

So at first, we are going to find the probability that the winning numbers of them are equal at least once in the process of competitions. The probability that East Wind wins totally x times in the competitions is

$$\frac{C_n^x}{2^n}.$$

According to what was discussed in the last chapter, for $x \leqslant \frac{n}{2}$ let $b = x$ and $a = n - x$, the probability that the winning numbers of them are equal at least once is then $\frac{2x}{n}$; for $x > \frac{n}{2}$ let $b = n - x$ and $a = x$, the corresponding probability is then $\frac{2(n-x)}{n}$.

Therefore, the probability that the winning numbers are equal at least once in the process of competitions is

$$\sum_{x\leqslant\frac{n}{2}} \frac{2x}{n} \cdot \frac{C_n^x}{2^n} + \sum_{x>\frac{n}{2}} \frac{2(n-x)}{n} \cdot \frac{C_n^x}{2^n}$$

$$= \frac{1}{2^{n-1}}\Big(\sum_{x\leqslant\frac{n}{2}} C_{n-1}^{x-1} + \sum_{x>\frac{n}{2}} C_{n-1}^{x}\Big) = \frac{1}{2^{n-1}}\Big(\sum_{x\leqslant\frac{n}{2}-1} C_{n-1}^{x} + \sum_{x>\frac{n}{2}} C_{n-1}^{x}\Big)$$

$$= \frac{1}{2^{n-1}}\Big(\sum_{x\leqslant n-1} C_{n-1}^{x} - C_{n-1}^{[\frac{n}{2}]}\Big) \left(\left[\frac{n}{2}\right] \text{ is the integer part of } \frac{n}{2}\right)$$

$$= \frac{1}{2^{n-1}}\left(2^{n-1} - C_{n-1}^{[\frac{n}{2}]}\right) = 1 - \frac{C_{n-1}^{[\frac{n}{2}]}}{2^{n-1}}.$$

Consequently, the probability that the winning numbers are not equal even once in the process of competitions is

$$\frac{C_{n-1}^{[\frac{n}{2}]}}{2^{n-1}}. \qquad (1)$$

The probability that East Wind overwhelms West Wind is then

$$\frac{1}{2} \times \frac{C_{n-1}^{[\frac{n}{2}]}}{2^{n-1}} = \frac{C_{n-1}^{[\frac{n}{2}]}}{2^{n}}. \qquad (2)$$

29. Dowry Problem (I)

Girls from Da Ban Town is a beautiful song that has been prevailing for 60 years in China. Some words in the song say: *With your dowry of millions, bring your young sister, and driving your cart, come to my house*.

Let us say, there are four girls from Da Ban Town, and one of them has a large dowry. The values of the dowries of these girls will be displayed on the screen one after one in a dark room, for a suitor to select. When the suitor enter the room, he will see the first number displayed on the screen; then he must decide to either select the first girl or give her up. If he decides to give up, he will see the second number displayed on the screen, and must decide to either select the second girl or give her up. If he still chooses to give up, he then will see the third number on the screen, and must decide to either select the third girl or give her up. If he still chooses to give up this time, then he must select the last girl.

The problem is: If the suitor hopes to select the girl with the largest dowry (i.e. the largest number), what strategy should he use?

Solution: As the suitor has not been told the values of the four dowries in advance, when seeing the first number he does not know if it is the largest number or not. So, if he selects this number, the probability that it is the largest is $\frac{1}{4}$.

Assume he gives up the first number. If he finds that the second number is smaller than the first one, he will of course give it up also. But when the second number is larger than the first, what should he do? He has two strategies to choose: One is to select the number that appears as soon as it is larger than the first; the other is to always give up the second.

For simplicity, we denote the four numbers, from small to large, by 1, 2, 3, and 4, respectively. Then there are $4! = 24$ permutations for them, as seen in the following.

1 234	2 134	3 124* △	4 123
1 243△	2 143* △	3 142△*	4 132
1 324△	2 314△	3 214* △	4 213
1 342△	2 341△	3 241* △	4 231
1 423*	2 413*	3 412*	4 312
1 432*	2 431*	3 421*	4 321

Among these permutations, there are 11 ones attached by an asterisk each, indicating the largest number can be selected on them by the first strategy. So the probability of selecting the girl with the largest dowry by this strategy is $\frac{11}{24}$.

In the case of choosing to always give up the second number, there are two further strategies: One is to select the number that appears as soon as it is larger than the first and second; the other one is to always give up the third (he then of course has to select the fourth). By the first strategy, there are 10 cases in the table above (attached by a triangle note each), on which the largest number can be selected. So the probability is $\frac{10}{24} = \frac{5}{12}$. By the latter strategy, the probability is only $\frac{1}{4}$ (as there are only 6 cases in the table that the number 4 is on the last place).

The results are summed as follows:

If he selects the first or the last, the probability is $\frac{1}{4}$.

If he gives up the first and select the number as soon as it is larger than the first, the probability is $\frac{11}{24}$.

If he gives up the first and second numbers and select the number as soon as it is larger than the previous two ones, the probability is $\frac{5}{12}$.

Therefore, the suitor should use the strategy "give up the first and select the number as soon as it is larger than the first".

If there are only two girls, the probability is obviously $\frac{1}{2}$. If there are three girls, then, by using the strategy recommended above, the probability is still $\frac{1}{2}$.

30. Dowry Problem (II)

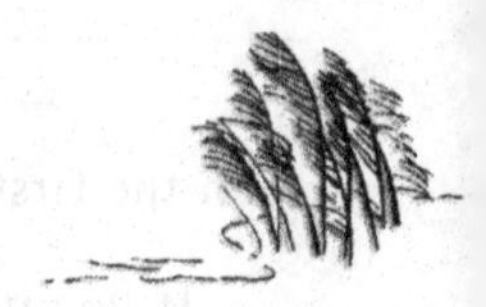

We continue the problem in the last chapter. Suppose the number of girls is generalized from 4 to $n(\geqslant 4)$. What strategy should the suitor use?

Solution: The way of solution is similar to that in the last chapter, but there are more strategies for the suitor to choose.

Suppose he gives up the first $s-1(s \geqslant 2)$ numbers, and select the number appeared later as soon as it is larger than all the previous ones. The probability that he gets the girl with the largest dowry by this strategy is denoted by p_s.

1. Find p_s

If the largest number m appears at the kth $(k \geqslant s)$ place in order, then the condition that m can be selected by the chosen strategy is: the largest number among the first $k-1$ ones appears in the first $s-1$ places. The occurrence probability of this condition is $\frac{s-1}{t-1}$ and that m appears at the kth place is $\frac{1}{n}$. Therefore,

$$p_s = \sum_{k=s}^{n}\left(\frac{1}{n} \times \frac{s-1}{k-1}\right) = \frac{s-1}{n}\sum_{k=s}^{n}\frac{1}{k-1}. \tag{1}$$

2. Compare p_s and p_{s+1}, to determine the largest p_s

$$p_s \geqslant p_{s+1} \Leftrightarrow (s-1)\sum_{k=s}^{n}\frac{1}{k-1} \geqslant s\sum_{k=s+1}^{n}\frac{1}{k-1}$$

$$\Leftrightarrow 1+\frac{s-1}{s}+\frac{s-1}{s+1}+\cdots+\frac{s-1}{n-1} \geqslant \frac{s}{s}+\frac{s}{s+1}+\cdots+\frac{s}{n-1}$$

$$\Leftrightarrow 1 \geqslant \frac{1}{s}+\frac{1}{s+1}+\cdots+\frac{1}{n-1}. \tag{2}$$

Therefore, when s satisfies

$$\frac{1}{s}+\frac{1}{s+1}+\cdots+\frac{1}{n-1}\leqslant 1<\frac{1}{s-1}+\frac{1}{s}+\cdots+\frac{1}{n-1}, \quad (3)$$

p_s is the largest.

For a given n, we can use (3) to determine s.

Example 1: When $n = 4$, we have

$$\frac{1}{2}+\frac{1}{3}<1<1+\frac{1}{2}+\frac{1}{3}.$$

Therefore $s = 2$, as is the result obtained in the last chapter.

Example 2: When $n = 5$, we have

$$\frac{1}{3}+\frac{1}{4}<1<\frac{1}{2}+\frac{1}{3}+\frac{1}{4}.$$

Therefore $s = 3$, which means the suitor should give up the first two numbers and select the number later as soon as it is larger than all the previous ones. The probability is then $p_s = \frac{2}{5}\left(\frac{1}{2}+\frac{1}{3}+\frac{1}{4}\right)=\frac{13}{30}$.

Example 3: When $n = 10$, we have

$$\frac{1}{4}+\frac{1}{5}+\cdots+\frac{1}{9}=0.999\,5\ldots.$$

So he should choose $s = 4$ (i. e. the first three numbers are abandoned).

When n is very large, it is necessary to use formula

$$1+\frac{1}{2}+\cdots+\frac{1}{n}=\log_e n+0.577\,21\ldots+\varepsilon_n, \quad (4)$$

where $e = 2.718\,28\ldots$ is the base of natural logarithm, $0.577\,21\ldots$ is called the Euler's constant, and ε_n is a very small number that could be neglected ($\varepsilon_n \to 0$, as $n \to \infty$).

From (4) we are able to obtain

$$p_s \approx \frac{1}{2}(\log_e(n-1)-\log_e(s-1))=\frac{s-1}{n}\log_e\frac{n-1}{s-1}.$$

By (3) and (4), we get

$$\frac{n-1}{s} \approx \mathrm{e} \approx \frac{n-1}{s-1}.$$

Consequently, when

$$s \approx \frac{n}{\mathrm{e}},$$

p_s becomes the largest, and

$$p_s \approx \frac{1}{\mathrm{e}} = 0.3678\ldots.$$

(We encounter $\frac{1}{\mathrm{e}}$ again! It has already appeared in Chapters 15 and 17.)

For example, when $n = 100$, we have $s = 38$ satisfying (3) and $p_s = 0.371$, which means the suitor should give up the first 37 numbers and select the number later as soon as it is larger than all of them. By this strategy, the probability that he gets the girl with the largest dowry is larger than $\frac{1}{3}$. By contrast, if he selects the first or the last girl, the corresponding probability is only $\frac{1}{100}$.

31. Job Interview

A company needs a secretary, and there are 10 people applying for the job. The manager of the company decides to have an interview with each of them in the time order of the applications. He will definitely deny the first three persons interviewed. Then he will deal with the following interviewees in this way: If the person during the interview shows a higher ability than all the previous interviewees, he will accept him/her; otherwise, he will deny him/her and proceed to the next interviewee. If he has denied the first 9 interviewees, he then has to accept the last one.

Suppose the abilities of these 10 people are different with each other and can be ranked from the highest to the lowest. Let A_k denote the number of those interview orders, by which the person ranked the kth place in ability will be accepted.

Please prove

(1) $A_1 > A_2 > \cdots > A_8 = A_9 = A_{10}$;

(2) the probability that one of the three persons with the highest abilities is accepted is more than 70%, and that one of the three persons with the lowest abilities is accepted will not exceed 10%.

Solution: (1) For convenience, the person with the ability ranked the kth ($2 \leqslant k \leqslant 10$) place is denoted as the k. For any interview order by which the k is accepted, we exchange the positions of the k and $k-1$ in it, and then get a new interview order by which the $k-1$ will be accepted instead. Therefore $A_{k-1} \geqslant A_k$. Furthermore, an order with the first four positions being $k+2$, $k+1$, k and $k-1$, respectively, will accept the $k-1$, but will not accept the k after the positions of the

k and $k-1$ are exchanged. This proves that $A_1 > A_2 > \cdots > A_8$. Finally, the 8, 9 and 10 will be accepted by an order only if they appear at the 10th place. At this time, if we, in an order accepting the $k-1(k=9$ or $10)$, exchange the positions of the k and $k-1$, the k will be accepted. Therefore $A_8 = A_9 = A_{10}$.

(2) If one of the 8, 9 and 10 is accepted by an interview order, he/she must be in the 10th place (with probability $\frac{3}{10}$) and the 1 in the first three positions (with probability $\frac{3}{9}$). Therefore, the total probability that one of the three persons with the lowest abilities is accepted is $\frac{3}{10} \times \frac{3}{9} = \frac{1}{10}$.

The probability that the 1 appears in the $k+1(3 \leqslant k \leqslant 9)$ th position is $\frac{1}{10}$, and the probability that the smallest number among the first k ones appears in the first three positions is $\frac{3}{k}$. Therefore, the probability that the 1 is accepted is

$$\frac{1}{10} \times \left(\frac{3}{3} + \frac{3}{4} + \frac{3}{5} + \frac{3}{6} + \frac{3}{7} + \frac{3}{8} + \frac{3}{9}\right) = 0.398\,690\,47\ldots.$$

The probability that the 2 appears at the $k+1(3 \leqslant k \leqslant 8)$ th place is $\frac{1}{10}$, and in the meantime, the probability that the 1 appears in the following $9-k$ positions is $\frac{9-k}{9}$; therefore the probability that the 2 is accepted in these positions is $\frac{1}{10} \times \frac{9-k}{9} \times \frac{3}{k}$. In addition, the probability that the 2 is at the 10th place is $\frac{1}{10}$, and in the meantime, the probability that the 1 is in the first three positions is $\frac{3}{9}$; therefore the probability that the 2 is accepted in this position is $\frac{1}{10} \times \frac{3}{9}$. In all, the total probability that the 2 is accepted is

$$\frac{1}{10}\times\frac{3}{9}\times\left(\frac{6}{3}+\frac{5}{4}+\frac{4}{5}+\frac{3}{6}+\frac{2}{7}+\frac{1}{8}+\frac{1}{1}\right)=0.198\,690\,47\ldots.$$

In a similar way, the probability that the 3 appears at the $k+1$ $(3\leqslant k\leqslant 8)$ th place is $\frac{1}{10}$, and in the meantime, the probability that the 1 and 2 appear in the following $9-k$ positions is $\frac{9-k}{9}\times\frac{8-k}{8}$; therefore the probability that the 3 is accepted in these positions is $\frac{1}{10}\times\frac{9-k}{9}\times\frac{8-k}{8}\times\frac{3}{k}$. In addition, the probability that the 3 is at the 10th place is $\frac{1}{10}$, and in the meantime, the probability that the 1 is in the first three positions is $\frac{3}{9}$; therefore the probability that the 3 is accepted in this position is $\frac{1}{10}\times\frac{3}{9}$. In all, the total probability that the 3 is accepted is

$$\frac{1}{10}\times\frac{3}{9}\times\left(\frac{6\times 5}{8\times 3}+\frac{5\times 4}{8\times 4}+\frac{4\times 3}{8\times 5}+\frac{3\times 2}{8\times 6}+\frac{2\times 1}{8\times 7}+\frac{1}{1}\right)$$
$$=0.111\,190\,47\ldots.$$

Finally, the probability that one of the three persons with the highest abilities (the 1, 2 and 3) is accepted is the sum of the three ones presented above:

$$0.398\,690\,47\ldots+0.198\,690\,47\ldots+0.111\,190\,47\ldots$$
$$=0.708\,5\ldots>70\%.$$

Note: The probability that the k is accepted is $\frac{A_k}{10!}$.

32. Boxing Match (I)

Boxers A and B agree to have a boxing match of an even number rounds between them. Unlike common boxing matches, this one will not sum the points of all rounds, but let every round have a winner; and the boxer who has won more than half the number of rounds will be the winner of the match.

Suppose the two boxers are well-matched in strength, i. e. the probabilities for them to win in each round is equally $\frac{1}{2}$. What are the probabilities that A wins the matches of 2 rounds, 4 rounds, and 6 rounds, respectively? How many rounds should the match have so that the probability for A to win it is the largest? And what is the largest probability for A to win the match?

Solution: If the match has only one round, the probability for A to win is then $\frac{1}{2}$.

But now only even numbers of rounds are allowed for the match.

If it is 2 rounds, the probability for A to win the match (i.e. win all the 2 rounds) is then $\frac{1}{2} \times \frac{1}{2} = \frac{1}{4}$.

If it is 4 rounds, the probability for A to win the match is

$$C_4^3 \times \left(\frac{1}{2}\right)^3 \left(\frac{1}{2}\right) + C_4^4 \times \left(\frac{1}{2}\right)^4 = \frac{5}{16}.$$

If it is 6 rounds, the probability for A to win is

$$(C_6^4 + C_6^5 + C_6^6) \times \left(\frac{1}{2}\right)^6 = \frac{22}{64} = \frac{11}{32}.$$

We can see from these cases that the probability for A to win increases as the number of rounds increases. As a matter of fact, assuming the match has $2n$ rounds, the probability that A wins is then

$$P_{2n} = \sum_{k=n+1}^{2n} C_{2n}^{k} \times \left(\frac{1}{2}\right)^{2n} = \frac{1}{2}\left[\sum_{k=n+1}^{2n} C_{2n}^{k} + \sum_{k=0}^{n-1} C_{2n}^{k}\right] \times \left(\frac{1}{2}\right)^{2n}$$

$$= \frac{1}{2}[(1+1)^{2n} - C_{2n}^{n}] \times \left(\frac{1}{2}\right)^{2n} = \frac{1}{2} - \frac{C_{2n}^{n}}{2} \times \left(\frac{1}{2}\right)^{2n}.$$

Since

$$\frac{C_{2n+2}^{n+1} \times \left(\frac{1}{2}\right)^{2n+2}}{C_{2n}^{n} \times \left(\frac{1}{2}\right)^{2n}} = \frac{2n+1}{2n+2} < 1,$$

$C_{2n}^{n} \times \left(\frac{1}{2}\right)^{2n}$ decreases as n increases. It is also provable that $C_{2n}^{n} \times \left(\frac{1}{2}\right)^{2n} \to 0$ as $n \to \infty$ (see the proof at the end of this chapter). Therefore, P_{2n} increases and tends to $\frac{1}{2}$ as n increases; that means the more the rounds, the larger the probability for A to win the match, and it comes nearer and nearer to $\frac{1}{2}$.

As A and B are well-matched in strength, so P_{2n}, the probability for A to win, is equal to Q_{2n}, that for B to win. Furthermore, we have

$$P_{2n} = Q_{2n} = \frac{1}{2}(1 - a_n),$$

where $a_n = C_{2n}^{n} \times \left(\frac{1}{2}\right)^{2n}$ is the probability that A and B have a draw in the match (i.e. each of them has won n rounds).

Therefore, when $a_n \to 0$, P_{2n} and $Q_{2n} \to \frac{1}{2}$.

It is interesting to see that the interests of A and B should be the same: Both of them wish that the number of rounds of the match be as large as possible, so that the probability for each of them to win is as great as possible coming nearer and nearer to $\frac{1}{2}$.

Finally, we prove that $a_n = C_{2n}^n \times \left(\frac{1}{2}\right)^{2n}$, as the coefficient of the middle term in the expansion of $(x+1)^{2n}$, tends to zero as $n \to \infty$. From

$$\begin{aligned} a_n &= \frac{(2n)!}{n!n!} \times \left(\frac{1}{2}\right)^{2n} \\ &= \frac{(2n-1)(2n-3) \times \cdots \times 3 \times 1}{(2n)(2n-2) \times \cdots \times 4 \times 2} \\ &< \frac{2n}{2n+1} \times \frac{2n-2}{2n-1} \times \cdots \times \frac{4}{5} \times \frac{2}{3} \\ &= b_n, \end{aligned}$$

we have

$$a_n^2 < a_n b_n = \frac{1}{2n+1}.$$

Then

$$a_n < \frac{1}{\sqrt{2n+1}}.$$

As $\frac{1}{\sqrt{2n+1}} \to 0(n \to +\infty)$, so $a_n \to 0(n \to 0)$.

Another way to prove is to use the famous Stirling's formula

$$n! = \sqrt{2n\pi}n^n e^{-n+\frac{\theta_n}{12n}} \ (0 < \theta_n < 1).$$

By it we have

$$\begin{aligned} a_n &= \frac{(2n)!}{n!n!} \times \left(\frac{1}{2}\right)^{2n} \\ &= \frac{(2n)^{2n}\sqrt{4n\pi}}{n^n n^n \times 2n\pi} \left(\frac{1}{2}\right)^{2n} \times e^{\varepsilon_n} \ (\varepsilon_n \to 0) \\ &= \frac{1}{\sqrt{n\pi}} \times e^{\varepsilon_n} \to 0. \end{aligned}$$

33. Boxing Match (II)

The rule of the boxing match is the same as that in the last chapter, except that the round number here m can be either odd or even. Suppose the probability for A to win each round is $p < \frac{1}{2}$. What m will make the probability for A to win the match the largest?

Solution: When $m = 1$, the probability for A to win the match is p.

Contrary to the last chapter, we are now going to prove that the probability for A to win is the largest when $m = 1$. When $m = 2n$, we have (let $q = 1 - p$)

$$p = p(p+q)^{2n-1} = p\sum_{k=0}^{2n-1} C_{2n-1}^{k} p^k q^{2n-1-k} = \sum_{k=1}^{2n} C_{2n-1}^{k-1} p^k q^{2n-k}$$

$$= \sum_{k=n+1}^{2n} C_{2n-1}^{k-1} p^k q^{2n-k} + \sum_{k=1}^{n} C_{2n-1}^{k-1} p^k q^{2n-k}$$

$$= \sum_{k=n+1}^{2n} C_{2n-1}^{k-1} p^k q^{2n-k} + \sum_{k=1}^{n} C_{2n-1}^{2n-k} p^k q^{2n-k}$$

$$= \sum_{k=n+1}^{2n} C_{2n-1}^{k-1} p^k q^{2n-k} + \sum_{k=n}^{2n-1} C_{2n-1}^{k} p^{2n-k} q^k$$

$$> \sum_{k=n+1}^{2n} C_{2n-1}^{k-1} p^k q^{2n-k} + \sum_{k=n+1}^{2n-1} C_{2n-1}^{k} p^{2n-k} q^k$$

$$> \sum_{k=n+1}^{2n} (C_{2n-1}^{k-1} + C_{2n-1}^{k}) p^k q^{2n-k} \;(\because p < q,\ k \geqslant 2n-k,$$

and we set $C_{2n-1}^{2n} = 0$)

$$= \sum_{k=n+1}^{2n} C_{2n}^{k} p^k q^{2n-k} = p_{2n}.$$

It means that the probability for A to win when m is an even number is less than p.

When $m = 2n+1$, we have

$$\begin{aligned}
p &= p(p+q)^{2n} = p\sum_{k=0}^{2n} C_{2n}^{k} p^{k} q^{2n-k} = \sum_{k=1}^{2n+1} C_{2n}^{k-1} p^{k} q^{2n+1-k} \\
&= \sum_{k=n+1}^{2n+1} C_{2n}^{k-1} p^{k} q^{2n+1-k} + \sum_{k=1}^{n} C_{2n}^{k-1} p^{k} q^{2n+1-k} \\
&= \sum_{k=n+1}^{2n+1} C_{2n}^{k-1} p^{k} q^{2n+1-k} + \sum_{k=n+1}^{n} C_{2n}^{k} p^{2n+1-k} q^{k} \\
&> \sum_{k=n+1}^{2n+1} (C_{2n}^{k-1} + C_{2n}^{k}) p^{k} q^{2n+1-k} \, (\because p < q,\ k \geqslant 2n+1-k) \\
&= \sum_{k=n+1}^{2n+1} C_{2n+1}^{k} p^{k} q^{2n+1-k} = p_{2n+1},
\end{aligned}$$

meaning the probability for A to win is still less than p even when m is odd.

We can, of course, also ask that: What m will make the probability for B to win the match the largest (i.e. $q_m = 1 - p_m$ reaches the largest)?

The reader might think this question over first. We will discuss it in the next chapter.

Remark: Zhou Wei-kang, a student from Jinling Middle School in Nanjing, proposed another proof, as shown in the following.

Suppose the match has m rounds, the probability for A to win the match is P and that for B is Q. Then

$$P + Q \leqslant 1, \tag{1}$$

where the both sides will be equal when m is odd, but does not when m is even (A and B may each win $\frac{m}{2}$ rounds with probability $C_{m}^{\frac{m}{2}} p^{m} q^{\frac{m}{2}} > 0$). We have

$$P = \sum_{k > \frac{m}{2}} C_{m}^{k} p^{k} q^{m-k}, \tag{2}$$

$$Q = \sum_{k>\frac{m}{2}} C_m^k p^{m-k} q^k. \tag{3}$$

Since $p < q$, then $\frac{q}{p} > 1$. When $k > \frac{m}{2}$, we have

$$\frac{C_m^k p^{m-k} q^k}{C_m^k p^k q^{m-k}} = \left(\frac{q}{p}\right)^{2k-m} \geqslant \frac{q}{p}.$$

That is

$$C_m^k p^{m-k} q^k \geqslant \frac{q}{p} \cdot C_m^k p^k q^{m-k}. \tag{4}$$

In (4), let $k = \left[\frac{m}{2}\right] + 1, \left[\frac{m}{2}\right] + 2, \ldots, m$, respectively, and sum them. We then have

$$Q \geqslant \frac{q}{p} P \tag{5}$$

and the both sides are equal only when $m = 1$.

Adding P to the both sides in (5) and combining (1), we get

$$1 \geqslant P + Q \geqslant \left(\frac{q}{p} + 1\right) P = \frac{P}{p}.$$

Therefore,

$$p \geqslant P,$$

where the both sides are equal only when $m = 1$.

34. Boxing Match (III)

Boxer A and B agree to have a match of an even number rounds between them, with the rule being the same as that in the last two chapters.

Suppose the probability for A to win each round is $p < \frac{1}{2}$. How many rounds do we need, so that the probability for A to win the match is the largest?

Solution: Assuming there are $2n$ rounds, then the probability for A to win the match is

$$P_{2n} = \sum_{k=n+1}^{2n} C_{2n}^{k} p^{k} q^{2n-k}, \tag{1}$$

where $q = 1 - p$. To let (1) be the largest, variable n should satisfy

$$P_{2n} \geqslant P_{2n+2}. \tag{2}$$

Although we can calculate $P_{2n} - P_{2n+2}$ directly, a simpler method is to analyze the differences between P_{2n} and P_{2n+2} at first: We see the part belonging to P_{2n} rather than P_{2n+2} is that A has won $n+1$ rounds in the first $2n$ ones but lost the last two rounds (the $(2n+1)$th and $(2n+2)$th), and what belonging to P_{2n+2} rather than P_{2n} is that A has won n rounds in the first $2n$ ones as well as won the last two rounds. Therefore, (2) becomes

$$q^{2} C_{2n}^{n+1} p^{n+1} q^{n-1} \geqslant p^{2} C_{2n}^{n} p^{n} q^{n}. \tag{3}$$

Simplifying it, we have

$$nq \geqslant (n+1)p. \tag{4}$$

Then we get

$$n \geqslant \frac{p}{q-p} = \frac{p}{1-2p}. \tag{5}$$

Therefore, when $n \geqslant \frac{p}{1-2p}$, we have $P_{2n} \geqslant P_{2n+2}$; but when $n < \frac{p}{1-2p}$, we have $P_{2n} \leqslant P_{2n+2}$. So when $n = \left\lceil \frac{p}{1-2p} \right\rceil$, P_{2n} is the largest. Here $\lceil x \rceil$ is called the ceiling function, representing the smallest integer not less than x.

For example, let $p \leqslant \frac{1}{3}$, $\frac{p}{1-2p} \leqslant 1$ and $\left\lceil \frac{p}{1-2p} \right\rceil = 1$. At this time, there should be 2 rounds, and the probability for A to win the match (i.e. win both rounds) is p^2.

Another example is $p = 0.45$. At this time there should be $2\left\lceil \frac{p}{1-2p} \right\rceil = 10$ rounds. It is not difficult to calculate the probability for A to win the match as

$$\begin{aligned}\sum_{k=6}^{10} C_{10}^{k} p^{k} q^{10-k} &\approx 0.159\,567\,755 + 0.074\,603\,106 + 0.022\,889\,589 \\ &\quad + 0.004\,161\,743 + 0.000\,340\,506 \\ &\approx 0.261\,562\,7.\end{aligned}$$

Notice that, when $n > \left\lceil \frac{p}{1-2p} \right\rceil$, P_{2n} is decreasing. Meanwhile $P_{2n} \geqslant 0$, so the sequence $\{P_{2n}\}$ has a definite limit. Then, what is it?

Since $p < \frac{1}{2}$ and $p + q = 1$, we have

$$4pq < (p+q)^2 = 1.$$

Then,

$$\begin{aligned}P_{2n} &= \sum_{k=n+1}^{2n} C_{2n}^{k} \times p^{k} q^{2n-k} < (4pq)^n \times \frac{1}{2^{2n}} \sum_{k=n+1}^{2n} C_{2n}^{k} \\ &< (4pq)^n \times \frac{1}{2^{2n}} \times \frac{1}{2}(1+1)^{2n} = \frac{1}{2}(4pq)^n \to 0 (n \to +\infty). \end{aligned} \tag{6}$$

Therefore, the limit of P_{2n} is zero.

On the other hand,

$$\begin{aligned} P_{2n-1} &= \sum_{k=n}^{2n-1} C_{2n-1}^{k} \times p^{k} q^{2n-1-k} = \sum_{k=n}^{2n-1} C_{2n-1}^{k} \times p^{k} q^{2n-1-k}(p+q) \\ &= \sum_{k=n}^{2n-1} C_{2n-1}^{k} \times p^{k+1} q^{2n-1-k} + \sum_{k=n}^{2n-1} C_{2n-1}^{k} \times p^{k} q^{2n-k} \\ &= \sum_{k=n+1}^{2n} C_{2n-1}^{k-1} \times p^{k} q^{2n-k} + \sum_{k=n}^{2n-1} C_{2n-1}^{k} \times p^{k} q^{2n-k} \\ &= C_{2n-1}^{n} p^{n} q^{n} + \sum_{k=n+1}^{2n} C_{2n}^{k} \times p^{k} q^{2n-k} = C_{2n-1}^{n} p^{n} q^{n} + P_{2n} \end{aligned} \tag{7}$$

and

$$C_{2n-1}^{n} p^{n} q^{n} = \frac{1}{2} C_{2n}^{n} p^{n} q^{n} < \frac{1}{2} C_{2n}^{n} \times \left(\frac{1}{4}\right)^{n} \to 0.$$

Therefore,

$$\lim_{n \to \infty} P_{2n-1} = \lim_{n \to \infty} P_{2n} = 0.$$

That is to say: P_m, the probability for A to win the match of m rounds, tends to zero as $m \to +\infty$.

Now, we are able to answer the question proposed at the end of the last chapter. Let the probability for B to win the match of m rounds be Q_m. Like what was discussed for (2), (3) and (4) (just exchange p and q), suppose

$$Q_{2n} \leqslant Q_{2n+2}. \tag{8}$$

It means (as $p < q$)

$$np \leqslant (n+1)q. \tag{9}$$

Therefore, Q_{2n} is increasing with m. Meanwhile $Q_{2n} \leqslant 1$, so the sequence $\{Q_{2n}\}$ has the limit as

$$\lim_{n \to \infty} Q_{2n} = 1 - \lim_{n \to \infty} P_{2n} = 1 - 0 = 1. \tag{10}$$

In a similar way, we can also prove that Q_{2n-1} is increasing with m:

$$
\begin{aligned}
Q_{2n+1} - Q_{2n-1} &= q^2 C_{2n-1}^{n-1} q^{n-1} p^n - p^2 C_{2n-1}^{n} q^n p^{n-1} \\
&= C_{2n-1}^{n-1} q^n p^n (q - p) \\
&> 0.
\end{aligned}
$$

And in a similar way as shown in (7) (just substitute P with Q, as well as exchange p and q), we have

$$Q_{2n-1} = Q_{2n} + C_{2n-1}^{n} p^n q^n.$$

Therefore,

$$\lim_{m \to \infty} Q_m = 1.$$

Then for B, the rounds of the match had better be an odd number, and the more the rounds the larger the probability for him to win the match. The probability will tend to 1 (i.e. he will certainly win).

35. Rein in on the Brink of the Precipice (I)

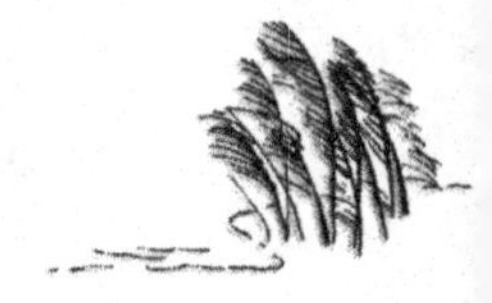

"A blind man riding a blind horse, midnight comes to the deep pool." — This old Chinese saying is a metaphor for a very dangerous situation.

Suppose a blind horse is on the brink of a precipice, and he will drop into it by just one step forward. Fortunately, the probability for him to do so is only $\frac{1}{3}$, while that for him to make one step backward is $\frac{2}{3}$. As the horse is wandering on the brink, what is the probability that he drops into the precipice?

Solution: Let the probability that the horse is at the place A, where it is one step away from the precipice, drops into it be P_1, and the probability that he is at place B, where it is two steps away from the precipice, drops into it be P_2. Then there are two ways for the horse at A to drop into the precipice.

(a) He makes one step forward, with probability $\frac{1}{3}$.

(b) He makes one step backward to reach B, but still drop into the precipice in the end with probability $\frac{2}{3}P_2$.

Therefore,

$$P_1 = \frac{1}{3} + \frac{2}{3}P_2. \tag{1}$$

On the other hand, the event that "the horse at B drops into the precipice" is the result of two events: "the horse reaches A" and "the horse at A drops into the precipice". These events are independent with each other and each has the occurrence probability P_1. So we have

$$P_2 = P_1^2. \tag{2}$$

Substituting (2) into (1), we get

$$P_1 = \frac{1}{3} + \frac{2}{3}P_1^2, \tag{3}$$

or

$$(P_1 - 1)(2P_1 - 1) = 0.$$

Then either $P_1 = 1$ or $P_1 = \frac{1}{2}$.

Whether $P_1 = 1$ or $P_1 = \frac{1}{2}$? It is a difficult question to answer.

We now consider a more general situation: the probability $\frac{1}{3}$ in the original problem is substituted by a real number $p\,(0 \leqslant p \leqslant 1)$. Obviously $P_1 = 0$ for $p = 0$, and $P_1 = 1$ for $p = 1$.

In a way similar to that illustrated above, we get

$$P_1 = p + (1-p)P_2, \tag{4}$$
$$P_2 = P_1^2 \tag{5}$$

and

$$P_1 = p + (1-p)P_1^2. \tag{6}$$

Therefore,

$$(P_1 - 1)\left(P_1 - \frac{p}{1-p}\right) = 0.$$

Then either $P_1 = 1$ or $P_1 = \frac{p}{1-p}$.

When $p = \frac{1}{2}$, we have $\frac{p}{1-p} = 1$. It means that Eq. (6) has only one solution $P_1 = 1$. So the blind man with the blind horse will certainly drop into the precipice. (What a pity!)

When $p > \frac{1}{2}$, the situation is of course still the worst $\left(\text{at this time } \frac{p}{1-p} > 1\text{, so it is impossible that } P_1 = \frac{p}{1-p}\right)$.

When $p < \frac{1}{2}$, we guess $P_1 = \frac{p}{1-p}$. It is true. In order to prove this, we compare the case where $p = \frac{1}{2}$ with that $p < \frac{1}{2}$ as follows.

The probability for the horse to drop into the precipice by one step forward is $\frac{1}{2}$ in the former case and is p in the latter one. Obviously $p < \frac{1}{2}$.

The probability that the horse drops into the precipice by first k steps backward and then $k+1$ steps forward is $\left(\frac{1}{2}\right)^{k+k+1}$ in the former case and in the latter one is $\left(\text{as } (1-p)p < \frac{1}{2} \times \frac{1}{2}\right)$

$$(1-p)^k p^{k+1} < \left(\frac{1}{2}\right)^k \left(\frac{1}{2}\right)^k \cdot p < \left(\frac{1}{2}\right)^{k+k+1}.$$

Therefore, the probability P_1 in the latter case is strictly less than that in the former one, which imply $P_1 = \frac{p}{1-p}$.

In summary,

$$P_1 = \begin{cases} 1 & \text{when } p \geqslant \frac{1}{2}, \\ \frac{p}{1-p} & \text{when } p < \frac{1}{2}. \end{cases}$$

In particular, $P_1 = \frac{1}{2}$ in the original problem $\left(\text{where } p = \frac{1}{3}\right)$.

(In Exercise 103 and its answer, there is another solution to this problem.)

36. Rein in on the Brink of the Precipice (II)

In general, suppose the horse is at the place M where it is m (a positive integer) steps away from the precipice, the probability he makes a step forward is p and that he makes a step backward is $1-p$.

Please find the probability P_m that the horse drops into the precipice.

Solution: We have already known from the last chapter that

$$P_1 = \begin{cases} 1 & \text{when } p \geqslant \frac{1}{2}, \\ \frac{p}{1-p} & \text{when } p < \frac{1}{2}. \end{cases}$$

From (2) in the last chapter, we have $P_2 = P_1^2$. Then

$$P_2 = \begin{cases} 1 & \text{when } p \geqslant \frac{1}{2}, \\ \left(\frac{p}{1-p}\right)^2 & \text{when } p < \frac{1}{2}. \end{cases}$$

Generally, the event "the horse is at M, m steps away from the precipice, and he finally drops into it" are the result of two events: "the horse is at M and he finally reaches A" and "the horse is at A, and he finally drops into the precipice". Therefore,

$$P_m = P_1 \times P_{m-1}. \tag{1}$$

Consequently,

$$P_m = P_1 \times P_{m-1} = P_1^2 \times P_{m-2} = \cdots = P_1^m = \begin{cases} 1 & \text{when } p \geqslant \frac{1}{2}, \\ \left(\frac{p}{1-p}\right)^m & \text{when } p < \frac{1}{2}. \end{cases}$$

37. Who Will Gamble Away?

Zhao has one yuan to gamble with and Qian has two yuan to do so. The bet in each round of the gambling between them is one yuan: the winner will take it from the loser, and there is no draw. The gamble will continue until one of them gambles away.

Suppose Zhao is a better gambler than Qian — the ratio of their winning chances in each round is two to one. Who is more possible to gamble away?

Solution: More generally, assume Zhao and Qian each has m yuan and n yuan to gamble with, respectively; the probability for Zhao to win in each round is p, and that for Qian is q, with $p + q = 1$ and $p > q$.

Suppose the probability for Zhao to gamble away is Q, and that for Qian is P.

We may regard Zhao as a horse at the place $x = m$ in the number axis, whose probability to move one step toward the origin is q (note that the q here is the p in the last chapter) and that to move one step in the opposite direction is p. According to what was discussed in the last chapter, the probability that the horse reaches the origin is $\left(\frac{q}{p}\right)^m$.

The event that the horse reaches the origin is a union of two ones: (a) it never reaches the place $x = m + n$ before, which means Zhao gambles away (with probability Q); and (b) it reaches the place first, meaning Qian has gambled away (with probability P), and then reaches the origin $\left(\text{with probability } \left(\frac{q}{p}\right)^{m+n}\right)$. Therefore,

$$\left(\frac{q}{p}\right)^m = Q + P \times \left(\frac{q}{p}\right)^{m+n}. \tag{1}$$

In a similar way, the probability that the horse reaches $x = m + n$ is 1. Therefore,

$$1 = P + Q \times 1. \tag{2}$$

(Besides the events that Zhao gambles away and that Qian gambles away, there is the third event: Neither of them will gamble away, which means the horse wanders between the origin and $x = m + n$, never reaching them. But the probability of this event is zero.)

Then we have

$$P\left(1 - \left(\frac{q}{p}\right)^{m+n}\right) = 1 - \left(\frac{q}{p}\right)^m, \tag{3}$$

$$P = \frac{1 - \left(\frac{q}{p}\right)^m}{1 - \left(\frac{q}{p}\right)^{m+n}}. \tag{4}$$

Therefore,

$$\begin{aligned} Q = 1 - P &= \frac{-\left(\frac{q}{p}\right)^{m+n} + \left(\frac{q}{p}\right)^m}{1 - \left(\frac{q}{p}\right)^{m+n}} \\ &= \left(\frac{q}{p}\right)^m \frac{1 - \left(\frac{q}{p}\right)^n}{1 - \left(\frac{q}{p}\right)^{m+n}} = \frac{1 - \left(\frac{p}{q}\right)^n}{1 - \left(\frac{p}{q}\right)^{m+n}}. \end{aligned} \tag{5}$$

For $m = 1$, $n = 2$, $p = \frac{2}{3}$ and $q = \frac{1}{3}$, we have

$$P = \frac{4}{7} > Q = \frac{3}{7}.$$

This result tells us that technique is more powerful than money, and Qian will gamble away more easily.

38. Equal in Strength

Continuing the problem in the last chapter, suppose Zhao and Qian each has m yuan and n yuan to gamble with, respectively, the bet for each round is one yuan, and the bet technique they possess are equal in power. Please find the probability that Zhao gambles away.

Solution: If $m = n$, then the conditions of them (money and technique) are the same; therefore, the probability for one of them to gamble away is the same as that for the other, i.e. $P = Q = \frac{1}{2}$.

If $m \neq n$, then by common sense, we know that the person who has more money is less possible to gamble away, as they are equal in the power of bet technique. Furthermore, it is reasonable to guess that the probability for Zhao to win is proportional to the money he has, i.e.

$$P = \frac{m}{m+n}. \tag{1}$$

In order to prove (1), we denote P as P_m. In general, if Zhao has j yuan to gamble with (Qian has $m + n - j$ yuan accordingly), the probability that he wins the gamble (equivalently Qian gambles away) is denoted as P_j.

When Zhao has j yuan to gamble with, then, after a bet round ends, there is a chance of $\frac{1}{2}$ to increase his money to $j + 1$ yuan, as well as the same probability to decrease his money to $j - 1$ yuan. Therefore,

$$P_j = \frac{1}{2}P_{j+1} + \frac{1}{2}P_{j-1}. \tag{2}$$

That means P_0, P_1, P_2, …, P constitute an arithmetical sequence. So

$$P_{m+n} = P_0 + (m+n)d, \tag{3}$$

where $d = P_1 - P_0$ is the common difference of the sequence.

Since $P_{m+n} = 1$ (at this time the money Qian has is zero) and $P_0 = 0$, then by (3) we get $d = \frac{1}{m+n}$. Consequently,

$$P = P_0 + md = \frac{m}{m+n}.$$

This completes the proof of (1).

Remark: Formula (4) in the last chapter cannot be used directly for $p = q$. However, if we let $p \to q$, then $\frac{q}{p} \to 1$; by using L'Hôpital's rule, we have

$$\lim_{p \to q} P = \lim_{x \to 1} \frac{1 - x^m}{1 - x^{m+n}} = \frac{m}{m+n}.$$

It is the same as (1).

39. Put All Money in One Bet

Gambler Zhao has 10 yuan, and want to earn another 10 yuan by playing roulette. It is said that the probability for a person to win in each round of roulette is 0.474 (less than $\frac{1}{2}$). If he puts a yuan bet in a round, then he may either win or lose a yuan by chance after the round. He can use either an aggressive strategy — putting all the 10 yuan in one bet, or a prudent one — putting 1 yuan in each bet. Which strategy will give him the larger probability to earn 10 yuan?

Solution: The probability for him to earn 10 yuan by the aggressive strategy is 0.474.

If he puts 1 yuan bet in each round, then the problem is the same as what discussed in Chapter 37: Under the condition that Qian also has 10 yuan to gamble with, find the probability for him to gamble away.

In formula (5) in Chapter 37, let $m = n = 10$, $p = 0.474$ and $q = 1 - p = 0.526$. The probability for Qian to gamble away is

$$P = \frac{1 - \left(\frac{0.526}{0.474}\right)^{10}}{1 - \left(\frac{0.526}{0.474}\right)^{20}} = 0.260\,97\ldots.$$

It means that the probability for Zhao to earn 10 yuan by the prudent strategy is only $0.260\,97\ldots < 0.474$. Therefore, he should use the aggressive strategy, to put all money in one bet.

40. Indeterminate Equation

There are red and black balls of the same size in a drawer. Suppose the probability for two balls taken randomly from the drawer being both red is $\frac{1}{2}$. At least how many balls are there in the drawer?

Solution: Assume there are totally t balls in the drawer, among which r ones are red. Obviously $t \geqslant r \geqslant 2$. As the probability of taking two red balls is $\frac{1}{2} < 1$, then $t \geqslant 3$.

The probability of taking two red balls is

$$\frac{r}{t} \times \frac{r-1}{t-1} = \frac{1}{2}. \tag{1}$$

From (1) we get

$$t^2 - t = 2(r^2 - r). \tag{2}$$

Equation (2) has two unknowns (t and r) and requires only positive integer solutions. Equations of this kind are called indeterminate equations or Diophantine equations.

In order to find integer solutions of (2), we at first change it, by multiplying 4 and adding 1 on both sides, into

$$(2t-1)^2 = 2(2r-1)^2 - 1, \tag{3}$$

and that is in the form of

$$x^2 - 2y^2 = -1, \tag{4}$$

with $x = 2t - 1$ and $y = 2r - 1$.

Equation (4) is one of the most common indeterminate equations, called Pell equation, whose solution has been studied in detail and can be found in related books. Here we only need to find the smallest positive integer x satisfying (4) by trying.

Since $t \geqslant 3$, we have $x = 2t - 1 \geqslant 5$. But $x = 5$ does not satisfy (4) ($(5^2 + 1) \div 2$ is not a square number). The next odd number is $x = 7$, and substituting it into (4) we get $y = 5$. Therefore $t = 4$ and $r = 3$ is the solution to our problem.

Remark: The minimum solution of (4) is $x_1 = 1$ and $y_1 = 1$; while $x_2 = 7$ and $y_2 = 5$ got above is the second minimum solution. Generally, all the solutions can be obtained by recursive formulas:

$$x_n = 2(2x_1^2 + 1)x_{n-1} - x_{n-2},$$

$$y_n = 2(2x_1^2 + 1)y_{n-1} - y_{n-2}.$$

For example, we can obtain from these formulas,

$$x_3 = 41, \ y_3 = 29; \ x_4 = 239, \ y_4 = 169.$$

41. Throw Copper Coin onto a Small Table

A number of masters of Kongfu (martial arts), including Guo Jing, Xiaoxiang Zi, and Jinlun Fawang, are coming together to have a special competition: A small square table in the tent from a distance that can be seen by all of them. Anyone who succeeds in throwing a copper coin onto the table (the diameter of the coin is $\frac{3}{4}$ the side of the square surface of the table) and keeping the entire coin within the boundary of the table surface will win the prize.

Suppose Guo Jing throws a copper coin onto the table successfully. Then what is the probability for him to win the prize?

Solution: This is a problem of geometric probability, which is a little different from that of the classical probability model that appeared in the previous chapters.

In a problem of geometric probability, the sample space I is a geometric figure. In our problem here, the sample space I is the square table surface, and an event is some part of I. The occurrence probability of the event is the ratio of the area of its part to that of I. (When I is a cube or a segment, the area should be changed to the volume or the length accordingly.)

We suppose, without loss of generality, the side of the square table surface is 1. A coin falls onto the table means equivalently that the center of the coin is within the surface.

The entire coin being within the surface means its center is in the small square of $\frac{1}{4} \times \frac{1}{4}$ around the center of the surface (see Fig. 1), so that the distance from it to any side of the surface is greater than $\frac{3}{8}$.

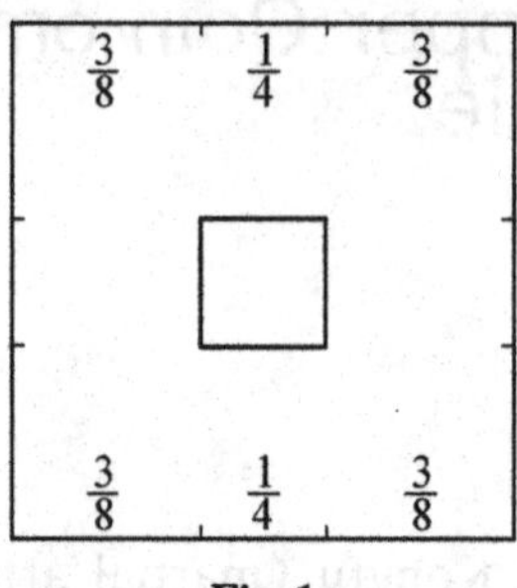

Fig. 1

As the area of the square table surface is 1 and that of the small square around the center of the surface is $\frac{1}{4} \times \frac{1}{4} = \frac{1}{16}$, the probability for Guo Jing to win the prize is then

$$\frac{\frac{1}{16}}{1} = \frac{1}{16}.$$

42. Appointment of People in Hurry

Wang has an appointment with Zhang between 9:00 and 10:00 tomorrow morning in front of the Oriental Pearl Tower. Both are busy persons, so each can stay there for only 5 minutes. Please find the probability that they are able to meet.

Solution: This is a problem of geometric probability again.

Let the time of Wang being at the place be $9 + x$ h, and that Zhang being there be $9 + y$ h. Then $0 \leqslant x \leqslant 1$ and $0 \leqslant y \leqslant 1$. Points (x, y) constitute a unit square in rectangular coordinate system with the vertexes (0, 0), (1, 0), (0, 1) and (1, 1) (see Fig. 1).

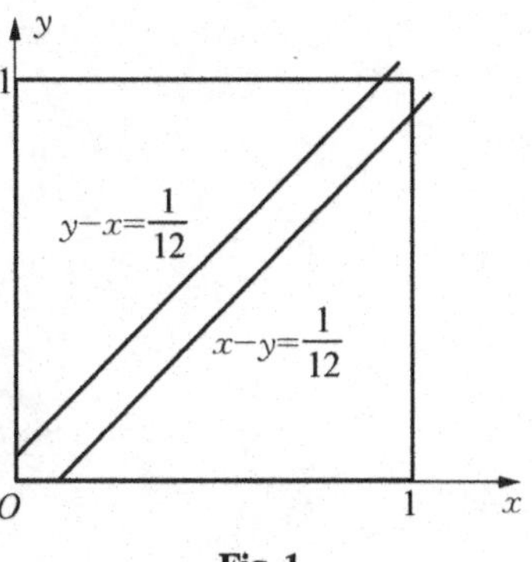

Fig. 1

As both persons stay just $\frac{1}{12}$ h (5 minutes), they can meet each other only when

$$|x - y| \leqslant \frac{1}{12}. \tag{1}$$

The graph of (1) is a zone in the square surrounded by two parallel lines $x - y = \frac{1}{12}$ and $y - x = \frac{1}{12}$, as is illustrated in Fig. 1. Outside the zone is two separated triangles, the sum of whose areas is

$$\left(1 - \frac{1}{12}\right) \times \left(1 - \frac{1}{12}\right) = \left(\frac{11}{12}\right)^2.$$

Then the area of the zone is

$$1-\left(\frac{11}{12}\right)^2=\frac{23}{144}.$$

Therefore, the required probability is

$$\frac{\frac{23}{144}}{1}=\frac{23}{144},$$

which is slightly less than $\frac{1}{6}$.

43. Obtuse Triangle

Taking randomly two positive numbers x and y, both smaller than 1, suppose they form a triangle with 1. Please find the probability that the triangle is obtuse.

Solution: This is also a problem of geometric probability.

As x, y and 1 form a triangle, we have

$$x + y > 1. \tag{1}$$

Therefore, the set of points (x, y) is the area of the triangle, $\triangle ABC$, in Fig. 1.

In addition, since the triangle formed by x, y and 1 is obtuse, then

$$x^2 + y^2 < 1. \tag{2}$$

The area of square $OABC$ is 1, that of $\triangle ABC$ is $\frac{1}{2}$, and that of the bow-shaped is $\frac{\pi}{4} - \frac{1}{2}$. So the required probability is

Fig. 1

$$\frac{\frac{\pi}{4} - \frac{1}{2}}{\frac{1}{2}} = \frac{\pi - 2}{2} \approx 0.570\,796\,3\ldots.$$

If x and y are just positive numbers less than 1, then the probability that they and 1 form an obtuse triangle is

$$\frac{\frac{\pi}{4}-\frac{1}{2}}{1}=\frac{\pi-2}{4}\approx 0.2853981\ldots.$$

We may ask: Under the condition that x, y and 1 form a triangle, what is the probability that the triangle is acute?

At this time, the set of points (x, y) is the area of $\triangle ABC$ minus the bow-shaped. Then the required probability is

$$1-\frac{\pi-2}{2}=\frac{4-\pi}{2}\approx 0.4292036\ldots,$$

which is less than that to form an obtuse triangle.

It is worthy to mention that the probability for x, y and 1 to form a right triangle is 0. Because the set of these points is the arch of the bow-shaped, whose area is 0. Then we see such kind of results: the probability of an event that may happen is 0. This phenomenon will not appear in the case where the sample space is a finite set (e. g. in the classical probability model).

44. Buffon's Needle

There are equal-distance parallel lines on a page of a usual exercise book. Suppose the distance is a. Now drop a needle of length $l(<a)$ onto the page randomly. Please find the probability that the needle lies across a line.

This problem was proposed by the famous French scholar Georges-Louis Leclerc, Comte de Buffon (1707 - 1788) in 1733 and solved by him in 1777.

Solution: Let the midpoint of the needle be M, the distance between it and the nearest parallel line be x, and the angle between the needle and the line be θ (see Fig. 1). Then we have

$$0 \leqslant x \leqslant \frac{a}{2} \text{ and } 0 \leqslant \theta \leqslant \pi. \tag{1}$$

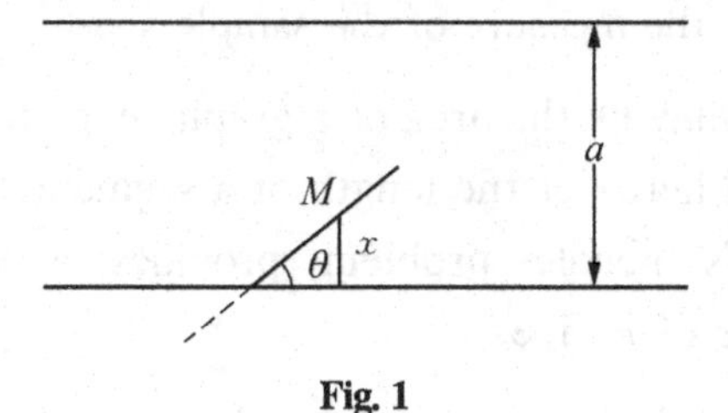

Fig. 1

If and only if

$$x \leqslant \frac{l}{2} \sin \theta, \tag{2}$$

the needle will be across the parallel line.

Expression (1) can be represented by a rectangular with (2) being the shade region in it (see Fig. 2).

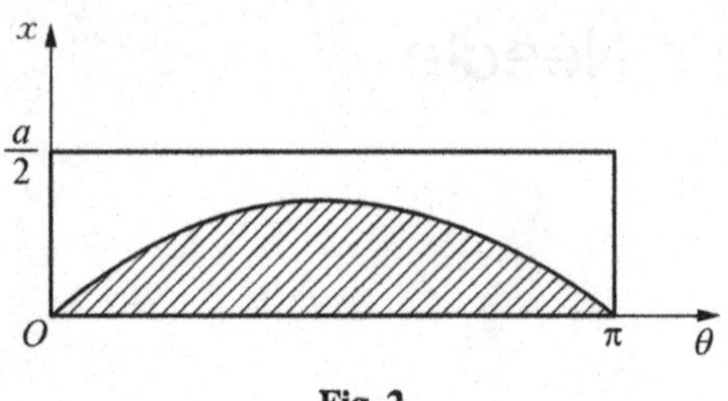

Fig. 2

The area of the rectangular is $\frac{a}{2}\pi$, and that of the shaded region is obtained by integral

$$\int_0^\pi \frac{l}{2}\sin\theta d\theta = \frac{l}{2}(-\cos\theta)\Big|_0^\pi = l.$$

The required probability is the ratio of these two areas, and that is

$$\frac{l}{\frac{a}{2}\pi} = \frac{2l}{a\pi}. \tag{3}$$

This is a classic problem of geometric probability. In a problem of this kind the probability of event A is

$$\frac{\text{the measure of } A}{\text{the measure of the sample space}},$$

where a measure may be the area of a graph (e.g. the Buffon's needle problem discussed here) or the length of a segment.

The Buffon's needle problem provides a way to find an approximate value of π, i.e.

$$\pi = \frac{a}{2l} \times \frac{\text{the number of times dropping the needle}}{\text{the number of times the needle lies across the line}}.$$

Now suppose the length of a needle $l \geqslant a$. What is the corresponding probability then?

At this time, when $\arcsin\frac{a}{l} \leqslant \theta \leqslant \pi - \arcsin\frac{a}{l}$, the needle will be across a parallel line surely; but when $0 \leqslant \theta \leqslant \arcsin\frac{a}{l}$ or $\pi -$

$\arcsin\frac{a}{l} \leqslant \theta \leqslant \pi$, the situation is the same as above, i.e. the needle is across a line when $x \leqslant \frac{l}{2}\sin\theta$. Then, the corresponding region is as shown in Fig. 3.

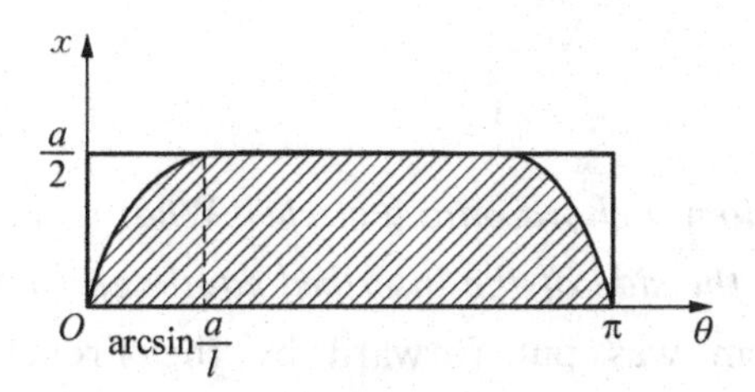

Fig. 3

The area of the shaded region can be calculated by integral

$$2\left(\int_0^{\arcsin\frac{a}{l}} \frac{l}{2}\sin\theta \mathrm{d}\theta + \frac{a}{2}\left(\frac{\pi}{2} - \arcsin\frac{a}{l}\right)\right)$$
$$= \frac{a}{2}\pi - a\arcsin\frac{a}{l} + l - l\cos\left(\arcsin\frac{a}{l}\right).$$

The required probability is then

$$1 - \frac{1}{\pi}\left(2\arcsin\frac{a}{l} - \frac{2l}{a} + \frac{2}{a}\sqrt{l^2 - a^2}\right).$$

When $l = a$, the corresponding probability is $\frac{2}{\pi}$. For a fixed a, let $l \to +\infty$. Then the probability tends to be 1.

The Buffon's Needle problem is an important mathematical model for mineral prospecting: Suppose there is a mineral vein of length l, and is prospected by a group of parallel lines. Then when $l < a$, the probability to find the mineral is $\frac{2l}{a\pi}$.

45. Bertrand's Paradox

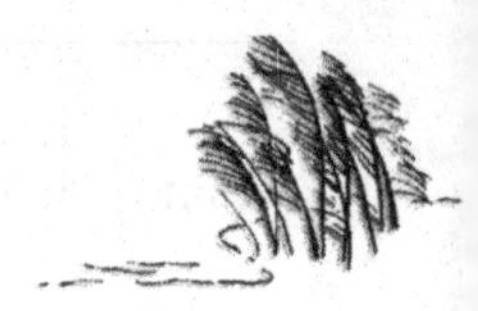

We draw at random a chord onto a circle. What is the probability that it is longer than the side of the inscribed equilateral triangle?

This problem was put forward by the French mathematician Joseph Louis François Bertrand (1822 – 1900) in his book *Calcul des probabilités* published in 1889. It has a number of solutions, and, quite an interesting thing, the results are not always the same.

Solution 1: We may assume that one end of the chord is a fixed point A (see Fig. 1). Suppose $\triangle ABC$ is the inscribed equilateral triangle. Then A, B and C trisect the circumference, and $AB = AC = \sqrt{3}$, if and only if point P on $\overset{\frown}{BC}$, the chord PA is longer than $\sqrt{3}$.

Therefore, the required probability is given by $= \dfrac{\text{the length of } \overset{\frown}{BC}}{\text{the length of the circumference}} = \dfrac{1}{3}$.

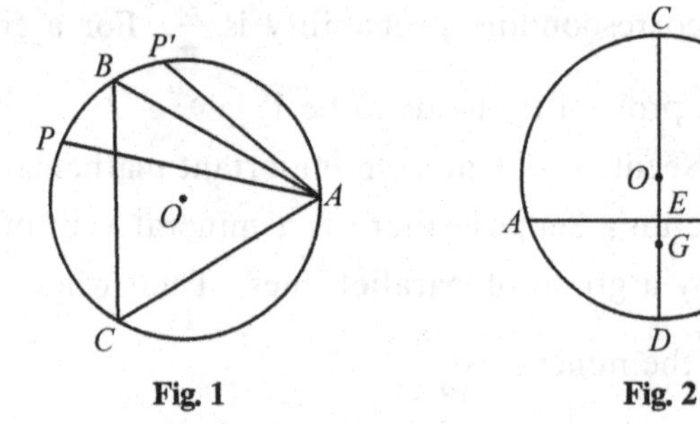

Fig. 1 Fig. 2

Solution 2: We may assume that the chord is perpendicular to a fixed diameter CD (see Fig. 2). Let O be the center of the circle, and G the midpoint of OD. If and only if the chord-center moment $OE < OG$, the chord PA is longer than $\sqrt{3}$. Therefore, the required probability is $\frac{OG}{OD} = \frac{1}{2}$.

Solution 3: Draw a concentric circle of $\odot O$ with diameter $\frac{1}{2}$, which is also the inscribed circle of the equilateral triangle (see Fig. 3). If and only if the midpoint E in the small circle, the chord PA is longer than $\sqrt{3}$. Therefore, the required probability is given by

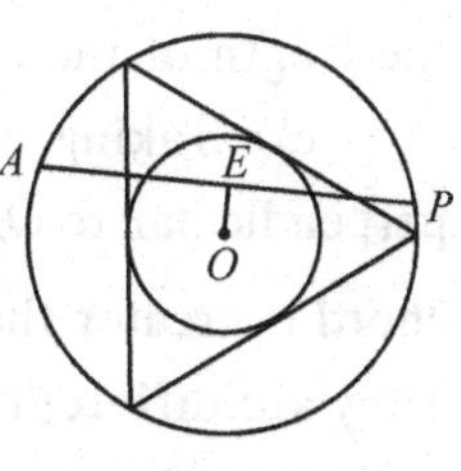

Fig. 3

$$\frac{\text{the area of the small circle}}{\text{the area of the large circle}} = \left(\frac{1}{2}\right)^2 = \frac{1}{4}.$$

The same problem should have the same answer. Here, however, we have three different answers, and each of them seems reasonable. It is quite strange!

What is wrong?

Firstly, the meaning of "draw at random a chord" is not clear. It is like to answer the question "How much time you need to go to school from home?" — If you can go in a random way, then, obviously, the answer will be different according to your choices: on foot, by bicycle or by bus. Now, the three solutions above are in fact the three answers to three different problems. They are:

(a) There is a fixed point A on the circumference. Now take a point P also on it. How much is the probability that the minor arc $\overset{\frown}{AP}$ is greater than one-third of the circumference?

(b) Taking a point E on a definite radius OD, please find the probability that $OE < \frac{1}{2}$.

(c) Taking any point P in $\odot O$, please find the probability that P is in a concentric cycle of $\odot O$ with the radius half of that of $\odot O$.

And they can also be expressed as:

(a′) There is a fixed point A on the circumference. Now take a point P also on it. How much is the probability that the length of $\overset{\frown}{AP}$ is greater than $\sqrt{3}$?

(b′) Taking a point E on a definite radius OD, and across E drawing a chord perpendicular to OD, please find the probability that the length of the chord is greater than $\sqrt{3}$.

(c′) Taking any point P in $\odot O$, and across P drawing a chord perpendicular to OP, please find the probability that the length of the chord is greater than $\sqrt{3}$.

They are different problems, and therefore have different answers.

Secondly, the sample spaces in the three problems are different from each other: The first one is the entire circumference, the second is a fixed radius, and the third the interior of $\odot O$. Correspondingly, the definitions of probability are also different from each other: The first one is the ratio of the length of an arc to that of the circumference, the second is the ratio of the length of a segment to that of the radius, and the third ratio is the area of a small circle to that of $\odot O$. In the case of the first definition, the equality of arc lengths implies the same probability; in that of the second definition, the equality of segment lengths implies the same probability; and in the case of the third, the equality of areas implies the same probability.

However, it is not difficult to see in Fig. 4: When $\overset{\frown}{AB} = \overset{\frown}{BC}$, we have the corresponding chords $AB = BC$; but their projections (onto the radius OD) A_1B_1 and B_1C_1 are not equal, as $\frac{A_1B_1}{B_1C_1} = \frac{\cos\alpha}{\cos\beta}$ where α and β are the angles made with OD by AB and BC, respectively. At the same time, the ring space between $\odot(O, OA_1)$, $\odot(O, OB_1)$ and that between $\odot(O, OB_1)$, $\odot(O, OC_1)$ are also not equal in areas, as the ratio of the areas of the two rings is

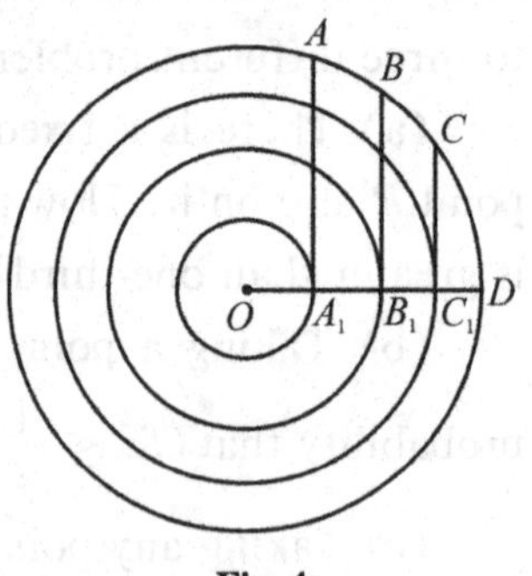

Fig. 4

$$\frac{A_1B_1 \times (OA_1 + OB_1)}{B_1C_1 \times (OB_1 + OC_1)}.$$

Even if $A_1B_1 = B_1C_1$, the areas of the two rings are still not equal. In all, the meanings of "equally possible distribution" (the same possibility) in the three cases are different from each other, so it is natural that their results are quite different. (Please recall the three statistics in Chapter 18.)

Bertrand's paradox reveals that: When a random experiment has infinite many possible results, it is necessary to have a clear definition of probability. This paradox has played an important role in the development of probability theory. In 1933, the Soviet mathematician Andrey Nikolaevich Kolmogorov (1903 - 1987) established an axiom system of probability theory for the first time in his book *Foundations of the Theory of Probability*, and let this theory become a rigorous branch of mathematics.

46. Odd or Even Number

Taking arbitrarily a number from the positive integers, what is the probability for it to be odd?

Solution: This problem seems quite easy: Since the odd and even numbers each occupy one half of all the positive integers, so the required probability is $\frac{1}{2}$.

If looking into it further, however, we would find that the problem is complicated: Why can we say "the odd and even numbers each occupies one half"? The reason seems that the set of positive integers can be divided into the set of positive odd integers and that of positive even ones, and the numbers of elements of two sets are equal.

But the set of positive odd integers and that of positive even ones are both infinite. When we say *the numbers of elements are equal* we mean there are one-to-one correspondences between the two sets. For example,

$$2n - 1 \leftrightarrow 2n \ (n = 1, 2, \ldots)$$

is one of this kind of correspondences.

On the other hand, we can also divide the set of positive integers into three subsets:

$$A = \{1, 5, 9, 13, \ldots\}, B = \{3, 7, 11, 15, \ldots\}$$

and the set of positive even integers. The numbers of elements of these three sets are also equal. For example,

$$4n - 3 \leftrightarrow 2n \ (n = 1, 2, 3, \ldots)$$

is a one-to-one correspondence between A and the set of positive even integers, and

$$4n-1 \leftrightarrow 2n \ (n=1,\ 2,\ 3,\ \ldots)$$

is a one-to-one correspondence between B and the set of positive even integers.

Accordingly, since the set of positive integers can be divided into three sets "with the same number of elements", and one of them is the set of positive even integers, then the probability of taking an even number should be $\frac{1}{3}$!

In a similar way, we can let the probability of taking an odd (even) number be $\frac{1}{k}$, with k being any natural number greater than 1.

It is something then like Bertrand's paradox.

The confusion is still caused by the unclear definition of probability (fail to define what an "equal possibility" is). There are a number of ways to save the situation. For example, we may at first consider taking a number from set $\{x \mid x \in \mathbf{N},\ x \leqslant M\}$ (where $\mathbf{N}$ is the set of all natural numbers, and M a positive real number). At this time, the set is finite and the chance of each number being taken is the same, so the probability to take an even number should be $\frac{1}{2}$ (when $[M]$ is even) or $\frac{[M]-1}{2[M]}$ (when $[M]$ is odd). Then let $M \to +\infty$ and get the limit $\frac{1}{2}$. We may define this limit as the required probability. By this definition, the probability to take an even number from the positive integers is $\frac{1}{2}$.

47. Rational or Irrational Number

Taking a number at random from interval [0, 1], what is the probability for it to be rational? And what is the probability for it to be irrational?

Solution: We should first make clear of the definition of probability, as we have already experienced things like this in the previous chapters.

The sample space here is interval [0, 1], an infinite set (be careful!) whose length is 1. For all the rational numbers in [0, 1], if the length formed by them is p, then

$$p = \frac{p}{1} \tag{1}$$

can be defined as the probability that "a number taken from [0, 1] is rational".

What is this p?

We can first list all the rational numbers (irreducible fractions) in [0, 1] one by one (like the positive integers can be listed as 1, 2, 3, ...):

$$0 = 1, 1 = \frac{1}{1}, \frac{1}{2}, \frac{1}{3}, \frac{2}{3}, \frac{1}{4}, \frac{1}{5}, \frac{3}{4}, \frac{2}{5}, \frac{1}{6}, \cdots \tag{2}$$

in the order of the sum of the numerator and denominator of each fraction, from small to large, and for those with the same sum (only a finite of them), the small denominator ones will be listed before.

Denote sequence (2) as $\{a_n\}$. For the nth term a_n in the sequence, we construct an interval $\left(a_n - \frac{\varepsilon}{2^{n+1}}, a_n + \frac{\varepsilon}{2^{n+1}}\right)$, where ε is an arbitrarily positive number. These intervals cover all the rational numbers in [0, 1]. Therefore, the mentioned p above is smaller than the sum of the lengths of them, which is

$$\frac{\varepsilon}{2}+\frac{\varepsilon}{4}+\cdots+\frac{\varepsilon}{2^n}+\cdots=\varepsilon\left(\frac{1}{2}+\frac{1}{4}+\cdots\right)=\varepsilon. \tag{3}$$

Then

$$p<\varepsilon. \tag{4}$$

We know from (4) that p is less than any positive number, so it must be zero.

Since the length (more accurately, the measure) of the rational numbers in [0, 1] is zero, then the required probability (i.e. that for a number taken randomly from [0, 1] to be rational) is 0.

Consequently, the probability for a number taken randomly from [0, 1] to be irrational is $1-0=1$.

Then we encounter again such kind of things: The probability for an event that may possibly happen is 0, while that for an event that may not happen certainly is 1.

The case above also tells us that: There are much more irrational number than rational ones in [0, 1]. Rational numbers are denumerable (can be listed as an infinite sequence like 1, 2, 3, 4, ...), while irrational numbers are not denumerable and the measure of them is equal to the whole sample space.

48. Real Roots or Not

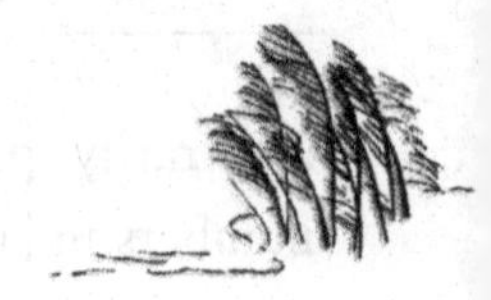

Let b and c be any real numbers. Please find the probabilities that the equation

$$x^2 + bx + c = 0 \qquad (1)$$

has and has not real roots, respectively.

Solution: Point (b, c) is distributed uniformly in the coordinate plane of infinite. We should adopt the method in Chapter 46, to look into the problem at first in a finite region — a square with the side $2B$, as seen in Fig. 1.

Fig. 1

The criterion of real roots of Eq. (1) is

$$b^2 \geqslant 4c. \qquad (2)$$

Here $b^2 = 4c$ is a parabola. In the square of the side $2B$ the area of the inside of the parabola (the shade region in Fig. 1) is

$$2 \times \frac{2}{3} \times B \times 2\sqrt{B}$$

(the area of curved line triangle OMN is one-third of that of rectangular $OMNB$ — a result obtained by Archimedes), from which we get the probability that Eq. (1) has no real roots in the square is

$$\frac{\frac{4}{3} \times B \times 2\sqrt{B}}{(2B)^2} = \frac{2}{3\sqrt{B}}.$$

As $B \to +\infty$ (the square being enlarged to the entire plane),

$$\frac{2}{3\sqrt{B}} \to 0.$$

Therefore, the probability that Eq. (1) has no real roots is zero, while that it has real roots is 1.

For more general quadratic equation

$$ax^2 + bx + c = 0 \ (a \neq 0), \tag{3}$$

we cannot transform it into

$$x^2 + \frac{b}{a}x + \frac{c}{a} = 0$$

first, and then discuss it (if we can, it will be reduced to the case above), because point (a, b, c) should be distributed uniformly in the three-dimensional coordinate system (except $a = 0$), and not be the case as $\left(\frac{b}{a}, \frac{c}{a}\right)$ is distributed uniformly in the coordinate plane.

The criterion of real roots of Eq. (3) is

$$b^2 \geqslant 4ac. \tag{4}$$

In the three-dimensional space, the equation

$$z^2 = 4xy \tag{5}$$

represents an elliptic cone. If you cannot see it, you may make rotation of $\frac{\pi}{4}$ for it, i.e. the coordinate transformation

$$x = \frac{\zeta + \eta}{\sqrt{2}}, \ y = \frac{\zeta - \eta}{\sqrt{2}},$$

to change (5) into

$$z^2 = 2(\zeta^2 - \eta^2), \tag{6}$$

or rewrite it as

$$\zeta^2 = \frac{z^2}{2} + \eta^2. \tag{7}$$

When ζ is constant k, (7) represents an ellipse (in plane $\zeta = k$); when $z = k\eta$ (k is a constant), (7) represents two straight lines (in plane $z = k\eta$) across the origin, as seen in Fig. 2.

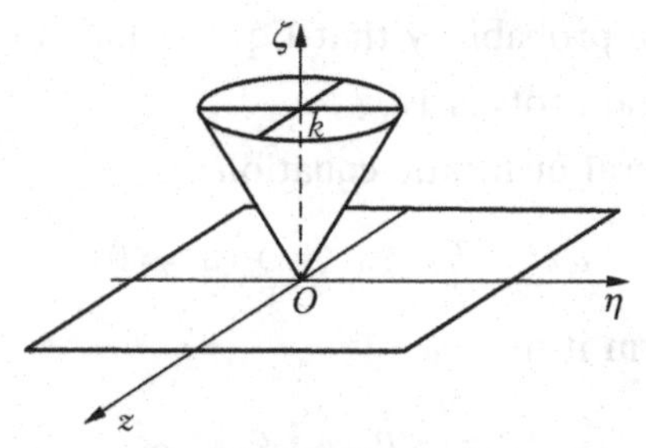

Fig. 2 (Only the part above plane $zO\eta$ is shown, while that below the plane is symmetric to it.)

Now, consider a cube of the side $2k$ centering O. The long axis of ellipse $\frac{z^2}{2} + \eta^2 = k^2$ is $\sqrt{2}k$ and its short axis is k, so its area is $\pi \cdot \sqrt{2}k^2$. Therefore, the volume of the elliptic cone in the cube is

$$2 \times \frac{1}{3}\pi \cdot \sqrt{2}k^2 \cdot k,$$

and the ratio of it to the cube is

$$\frac{2 \times \frac{1}{3}\pi \cdot \sqrt{2}k^3}{(2k)^3} = \frac{\sqrt{2}\,\pi}{12} = 0.370\,240\,2\ldots.$$

That is to say the probability that Eq. (3) has no real roots is $0.370\,240\,2\ldots$, while that it has real roots is $0.629\,759\,7\ldots$.

49. Divide the Stake

Two gamblers A and B agree that whoever wins s rounds of gambling first will take the entire stake. However, when A and B have won a and b $(a, b < s)$ rounds, respectively, the gambling is interrupted (by unknown reason, or maybe by a police man to catch them). Please find a reasonable way to divide the stake.

Remark: This is one of the most famous problems in the history of probability theory. In 1651, a well-known gambler de Méré (Antoine Gombaud, Chevalier de Méré, 1607 – 1684) proposed this problem to the distinguished French mathematician Blaise Pascal (1623 – 1662) for solution. For de Méré, this is a practical problem: Someone had suggested to divide the stake by ratio $a : b$, but he thought it unreasonable, because it fails to consider the situation that one of the gamblers might win the stake quickly if the gambling were going on (e.g. when $a = s - 1$, A might get the stake by only one round). de Méré is not only a gambler but also a statistician, so his argument really makes the point. Pascal discussed the problem with another distinguished French mathematician Pierre de Fermat (1601 – 1665). Later, Dutch mathematician and scientist Christiaan Huygens (1629 – 1695) joined the discussion. They all had found the correct solution. The content of the discussion was recorded in Huygens' Book *De ratiociniis in ludo aleae* ("On Reasoning in Games of Chance", 1657).

Solution: Let the probability for A to win each round be p $(0 < p < 1)$, and that for B be $q = 1 - p$; $m = s - a$ and $n = s - b$. Then only $m + n - 1$ additional rounds are needed to determine who will win the stake.

For A, he only needs to win m or more rounds. Therefore, the probability for him to win the stake is

$$P_1 = \sum_{i=m}^{m+n-1} C_{m+n-1}^{i} p^i q^{m+n-1-i}.$$

In a similar way, that for B is

$$P_2 = \sum_{i=n}^{m+n-1} C_{m+n-1}^{i} p^{m+n-1-i} q^i.$$

Accordingly, the stake should be divided by ratio $P_1 : P_2$.

It is not difficult to see that $P_1 + P_2 = 1$.

Some people take 29th July 1654 (the day Pascal wrote to Fermat to discuss the problem) as the birth day of probability theory. Some people also say that the probability theory originates from gambling. Although these arguments are imperfect, they are, however, not absolutely groundless.

On the other hand, although probability theory has something to do with gambling and there are also several examples concerning gambling in this book, these examples can also be applied to other cases, such as the motion of gas molecules or particles, the predication of economy or military affairs, weather forecast, etc. Gambling is only a kind of model where gamblers can be explained as a pair of opponents in various situations.

From the problem of dividing the stake we derive an important concept in the probability theory: *mathematical expectation*, or briefly *expectation*.

Suppose the stake is M yuan, the probability for A to win (i. e. get the entire stake) is p, and that for him to lose (i. e. get 0 yuan) is $q = 1 - p$. Then the mathematical expectation of A is pM (i. e. the stake should be divided by ratio $p : q$).

More generally, if the range of a random variable X is x_1, x_2, ..., x_n and the probability for it to have value x_k is p_k $(1 \leqslant k \leqslant n)$ with $p_1 + p_2 + \cdots + p_n = 1$, then

$$\sum_{k=1}^{n} x_k = x_1 p_1 + x_2 p_2 + \cdots + x_n p_n \tag{1}$$

is called the mathematical expectation of X, denoted as $E(X)$.

In the problem of dividing the stake discussed above, $n = 2$, $x_1 = M$, $x_2 = 0$, $p_1 = p$, $p_2 = 1 - p = q$ and $E(X) = Mp$. When $M = 1$, we have $E(X) = p$, meaning the mathematical expectation at this time is the probability.

The mathematical expectation is also called an average value, as it is a kind of weighted averages.

Suppose the assignment score of a student is 80, his midterm test score is 85 and terminal examination score is 90. If the assignment score, midterm test score and terminal examination score each occupies 30%, 30% and 40% of the final score, respectively, then his final score is

$$80 \times 30\% + 85 \times 30\% + 90 \times 40\% = 85.5.$$

Here the scores 80, 85 and 90 are respectively x_1, x_2 and x_3 ($n = 3$) in (1), while 30%, 30% and 40% corresponding to p_1, p_2 and p_3, respectively. These probabilities are also frequently called weights.

50. Sleeping Beauty

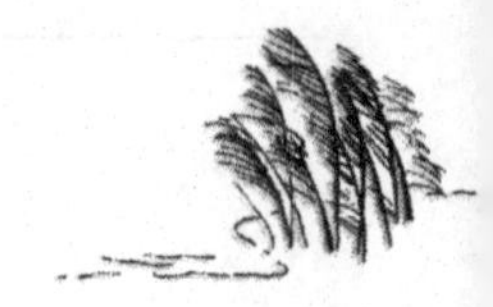

A beautiful and highly gifted princess has been confined in a castle. Now there are n keys distributed to n heroes, respectively; the heroes will try to open the door of the castle with their keys in turn, and among the keys only one matches the door. We ask: Which hero has the largest probability to open the door? And how many tries in average are needed to open the door?

Solution: Many people believe that earlier heroes may have larger chances (probabilities) to open the door. It is actually, however, not so.

The probability for the first hero to open the door is $\frac{1}{n}$.

The probability for the second hero to open the door is

$$\left(1-\frac{1}{n}\right)\times\frac{1}{n-1}=\frac{1}{n}$$

(where $1-\frac{1}{n}$ is the probability that the first hero fails to open the door, and then $\frac{1}{n-1}$ is that for the second one to succeed), and so on.

The probability for the kth hero to open the door is

$$\left(1-\frac{1}{n}\right)\left(1-\frac{1}{n-1}\right)\cdots\left(1-\frac{1}{n-k+2}\right)\times\frac{1}{n-k+1}=\frac{1}{n}$$

($1-\frac{1}{n}$ is the probability for the first hero to fail, $1-\frac{1}{n-1}$ is that for the second hero to fail after the first one and so on, and $\frac{1}{n-k+1}$ is the probability for the kth hero to succeed after the first $k-1$ fails), and so on.

The probability for the nth hero to open the door is

$$\left(1-\frac{1}{n}\right)\left(1-\frac{1}{n-1}\right)\cdots\left(1-\frac{1}{2}\right)\times 1=\frac{1}{n}.$$

Therefore, the probability for each hero (no matter which place he occupies in the order of trying) is the same.

The probability to open the door by k tries (i.e. the kth hero has succeeded) is $\frac{1}{n}$ ($k=1, 2, \ldots, n$). Therefore, the number of average tries needed to open the door is

$$1\times\frac{1}{n}+2\times\frac{1}{n}+\cdots+n\times\frac{1}{n}=\frac{1}{n}(1+2+\cdots+n)=\frac{n+1}{2}.$$

51. Number of Hits

Shoot a target n times in a row. It is known that the probability to hit the target each time is p. What number of hits in the n shootings can be expected?

Solution: According to Chapter 10, the probability of hitting k times in n shootings is

$$C_n^k p^k q^{n-k} \ (q = 1 - p) \tag{1}$$

(here we assume that the n shootings are independent with each other; otherwise, if it is allowed to make adjustment after each shooting, then the probability of hit will increase gradually, not always being p). Then the expectation of the number of hits is

$$\sum_{k=0}^{n} k C_n^k p^k q^{n-k}. \tag{2}$$

In order to calculate this sum, we need the expression (when $k \geqslant 1$)

$$k C_n^k = k \times \frac{n}{k} C_{n-1}^{k-1} = n C_{n-1}^{k-1}.$$

Then we have

$$\begin{aligned}
\sum_{k=0}^{n} k C_n^k p^k q^{n-k} &= \sum_{k=1}^{n} k C_n^k p^k q^{n-k} = \sum_{k=1}^{n} n C_{n-1}^{k-1} p^k q^{n-k} \\
&= np \sum_{k=1}^{n} C_{n-1}^{k-1} p^{k-1} q^{n-k} \\
&= np \sum_{k=0}^{n-1} C_{n-1}^{k-1} p^k q^{n-k-1} \text{(replace } k-1 \text{ by } k\text{)} \\
&= np\,(p+q)^{n-1} = np.
\end{aligned}$$

And that means we can expect to hit the target np times (if the answer is required an integer, then $[np]$ can be used instead).

Another way of solution is: Define the number of hits as X, and that in the ith shooting is $X_i (1 \leqslant i \leqslant n)$. Then

$$X_i = \begin{cases} 1 & \text{with probability } p, \\ 0 & \text{with probability } q, \end{cases} (1 \leqslant i \leqslant n),$$

and

$$X = X_1 + X_2 + \cdots + X_n.$$

Therefore,

$$E(X) = E(X_1) + E(X_2) + \cdots + E(X_n).$$

As it is obvious that $E(X_1) = E(X_2) = \cdots = E(X_n) = p$. So $E(X) = np$.

52. The Suicide Club

New Arabian Nights is a collection of short stories written by the well-known Scottish novelist Robert Louis Stevenson (1850 - 1894), in which he told a horrible story, *The Suicide Club*: Every evening, the president of the club would deal cards to each member, and anyone who got the spade ace must commit to "suicide" (actually, murdered by the president).

Putting aside the dirty plot in it, we now discuss such a probability problem: Suppose deal cards one by one, till the spade ace appears. How many cards will be dealt?

Solution: Assume that k cards have been dealt before the spade ace appears. Then $0 \leqslant k \leqslant 51$ (two jokers discarded, there are 52 cards).

The required probability for $k = 0$ is $\frac{1}{52}$.

The required probability for $k = 1$ is $\frac{51}{52} \times \frac{1}{51} = \frac{1}{52}$.

$\vdots$

The required probability for $k = 51$ is $\frac{51}{52} \times \frac{50}{51} \times \cdots \times \frac{1}{2} \times 1 = \frac{1}{52}$.

(As a matter of fact, "the spade ace" here is "the sleeping beauty" in the last chapter. We may also consider the problem from another perspective: The spade ace can be placed before or after any other 51 cards, and there are 52 available places for it with equal chance, and that is $\frac{1}{52}$.)

The average number of cards is then

$$\sum_{k=0}^{51} \frac{1}{52} \times k = \frac{1}{52} \times \frac{51(51+1)}{2} = \frac{51}{2} = 25.5,$$

and that means 25.5 cards in average have been dealt before the spade ace appears. So

$$25.5 + 1 = 26.5$$

cards in average have been dealt, till the appearance of the spade ace.

A simpler way to solve the problem is: We place the spade ace first, and then place each of the other cards either before or after it with the same probability $\frac{1}{2}$. Therefore, the average number of cards before the spade ace should be

$$51 \times \frac{1}{2} = 25.5.$$

The former solution way starts from the number of cards, while the latter one starts from each card. They are equivalent to the two solution ways described in the last chapter (except that we have not written explicitly the random variables in the latter case).

53. The First Ace

Suppose "the suicide club" has revised its rule, stipulating: Anyone who get the first ace (no matter which suit it belongs to) must commit to suicide. How many cards in average will be dealt before the first ace appears?

Solution: Assume that $k\,(0 \leqslant k \leqslant 48)$ cards have been dealt before the first ace appears. Then

the required probability for $k = 0$ is $\frac{4}{52}$,

the required probability for $k = 1$ is $\frac{48}{52} \times \frac{4}{51}$,

⋮

the required probability for $k = 48$ is $\frac{48}{52} \times \frac{47}{51} \times \cdots \times \frac{1}{5} \times \frac{4}{4}$.

The sum of them is 1 (as the appearance of an ace is a certain event), so

$$1 = \frac{4}{52} + \sum_{k=1}^{48} \frac{4}{52} \times \frac{48}{51} \times \frac{47}{50} \times \cdots \times \frac{49-k}{52-k}. \qquad (1)$$

The required average number (i.e. the mathematical expectation) is

$$A = \sum_{k=0}^{48} k \times \frac{4}{52} \times \frac{48}{51} \times \frac{47}{50} \times \cdots \times \frac{49-k}{52-k}. \qquad (2)$$

By Eq. (1) $\times 52 -$ (2), we get

$$52 - A = 4 + \sum_{k=1}^{48} \frac{4}{52} \times \frac{48}{51} \times \cdots \times \frac{50-k}{53-k} \times (49-k).$$

That is

$$A = 48 - \frac{4}{52} \times 48 - \frac{4}{52} \times \frac{48}{51} \times 47 - \cdots - \frac{4 \times 48 \times 47 \times \cdots \times 2}{52 \times 51 \times \cdots \times 5} \times 1$$
$$= 48 \times \left(1 - \frac{4}{52} - \frac{47 \times 4}{52 \times 51} - \cdots - \frac{47 \times 46 \times \cdots \times 1}{52 \times 51 \times \cdots \times 6} \times \frac{4}{5}\right). \tag{3}$$

Note that

$$\frac{4}{5} - \frac{4}{52} - \frac{47 \times 4}{52 \times 51} - \cdots - \frac{47 \times 46 \times \cdots \times 1}{52 \times 51 \times \cdots \times 6} \times \frac{4}{5}$$
$$= \frac{47 \times 4}{52 \times 5} - \frac{47 \times 4}{52 \times 51} - \cdots - \frac{47 \times 46 \times \cdots \times 1}{52 \times 51 \times \cdots \times 6} \times \frac{4}{5}$$
$$= \frac{47 \times 46 \times 4}{52 \times 51 \times 5} - \cdots - \frac{47 \times 46 \times \cdots \times 1}{52 \times 51 \times \cdots \times 6} \times \frac{4}{5}$$
$$\vdots$$
$$= \frac{47 \times 46 \times \cdots \times 2 \times 4}{52 \times 51 \times \cdots \times 7 \times 5} - \frac{47 \times 46 \times \cdots \times 2 \times 4}{52 \times 51 \times \cdots \times 7 \times 6} - \frac{47 \times 46 \times \cdots \times 1}{52 \times 51 \times \cdots \times 6} \times \frac{4}{5}$$
$$= \frac{47 \times 46 \times \cdots \times 2 \times 1 \times 4}{52 \times 51 \times \cdots \times 7 \times 6 \times 5} - \frac{47 \times 46 \times \cdots \times 1}{52 \times 51 \times \cdots \times 6} \times \frac{4}{5} = 0.$$

Therefore, $A = 48 \times \left(1 - \frac{4}{5}\right) = \frac{48}{5} = 9.6$, i.e. 9.6 cards in average have been dealt before the appearance of the first ace.

The calculating skill shown above deserves our attention.

On the other hand, similar to the last chapter, we may look into the problem from other perspective to simplify it: We at first place the four aces in a row, so that five intervals are formed either before or after each of them. The other 48 cards are then placed randomly in these intervals with the same probability $\frac{1}{5}$. Therefore, there are

$$48 \times \frac{1}{5} = 9.6$$

cards in average to appear before the first ace.

54. How Many Pairs in Average

There are two packs of poker cards. We at first place the first pack of cards (52 in all) in a row, from left to right, and then place the second pack on the first ones, respectively, to form 52 card couples. How many pairs (i.e. those consisting of the same card) are there in average among these card couples?

Solution: For each card in the second pack, there are 52 available positions, among which only one matches it, and that means its probability is $\frac{1}{52}$. Therefore, the average match value (the mathematical expectation) for each card in the second pack is

$$1 \times \frac{1}{52} + 0 \times \frac{51}{52} = \frac{1}{52}.$$

In sum, the average match value of all the 50 cards in the second pack is

$$52 \times \frac{1}{52} = 1.$$

So the average number of pairs is 1.

If two jokers are included to become a pack of 54 poker cards, the number of pair in average is still 1.

In fact, no matter what the number n of cards in each of the two same packs is, the number of pairs of the two same card is always

$$n \times \frac{1}{n} = 1.$$

In a similar way, if someone has written n letters and n envelopes, and his grandson puts randomly each letter into each envelope, then the average number of correct pairs is only 1.

Furthermore, in Chapter 17 we have obtained the probability of having exactly r correct pairs is

$$\frac{1}{r!}\left(\frac{1}{2!}-\frac{1}{3!}+\cdots+\frac{(-1)^k}{k!}+\cdots+\frac{(-1)^{n-r}}{(n-r)!}\right).$$

The mathematical expectation of the number of correct pairs is

$$\sum_{r=1}^{n} r\times\frac{1}{r!}\left(\frac{1}{2!}-\frac{1}{3!}+\cdots+\frac{(-1)^k}{k!}+\cdots+\frac{(-1)^{n-r}}{(n-r)!}\right)$$
$$=\sum_{r=1}^{n}\frac{1}{(r-1)!}\sum_{k=0}^{n-r}\frac{(-1)^k}{k!}.$$

Comparing it with the former result, we derive an interesting identical equation:

$$1=\sum_{r=1}^{n}\frac{1}{(r-1)!}\sum_{k=0}^{n-r}\frac{(-1)^k}{k!}. \tag{1}$$

Identity (1) can be proved directly as follows:

$$\begin{aligned}\text{The right side of (1)} &= \sum_{r=1}^{n}\sum_{k=0}^{n-r}\frac{(-1)^k}{(r-1)!k!}\\ &= \sum_{t=1}^{n}\sum_{r=1}^{t}\frac{(-1)^{t-r}}{(r-1)!(t-r)!}\quad (\text{let } k+r=t)\\ &= \sum_{t=1}^{n}\frac{(-1)^{t-1}}{(t-1)!}\sum_{r=1}^{t}\frac{(-1)^{r-1}\cdot(t-1)!}{(r-1)!(t-r)!}\\ &= \sum_{t=1}^{n}\frac{(-1)^{t-1}}{(t-1)!}\sum_{s=0}^{t-1}(-1)^s C_{t-1}^{s}.\end{aligned}$$

As $\sum_{s=0}^{t-1}(-1)^s C_{t-1}^{s}=0$ for $t>1$, we get that the right side of (1) is equal to 1. This completes the proof.

To prove an identical equation by using methods from the probability theory, such a kind of things we have already met before (e.g. in Chapter 19).

55. Many Holidays

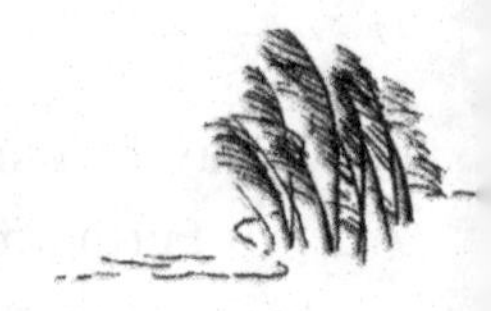

Republic of Roska practices welfarism, stipulating: A factory should give all his workers the entire day off, as long as it is the birthday of any worker in the factory. Therefore, if a factory has only one worker, then there are in fact 364 man-days in a year; if it has two worker, there may be 2×363 man-days yearly. In order to expect the largest number of man-days in a year, how many workers should a factory have? (Assume that there are 365 days in a year.)

Solution: Suppose there are n workers in a given factory. If all the workers have the same birthday, then, of course, the more workers there are, the more man-days it will have. But if everyday is the birthday of some worker, then the factory cannot operate. So more workers may not ensure more man-days.

The probability for a day not to be the birthday of a given worker is $\frac{364}{365}$, and then that for the day not be the birthday of all the n workers is $\left(\frac{364}{365}\right)^n$ — It is also the probability that the factory operates in the day. Therefore, the expected number of man-days in a year is

$$n \times 365 \times \left(\frac{364}{365}\right)^n. \tag{1}$$

To find n that makes (1) the largest, we solve inequality

$$(n+1) \times 365 \times \left(\frac{364}{365}\right)^{n+1} \leqslant n \times 365 \times \left(\frac{364}{365}\right)^n, \tag{2}$$

and get

$$\frac{(n+1)}{n} \leqslant \frac{364}{365}.$$

Then $n \geqslant 364$, and the equality in (2) holds when $n = 364$.

Therefore, when there are 364 or 365 workers, the expected number of man-days reaches the largest.

56. Buy Lottery Tickets

You buy a lottery ticket with one yuan, and then have a chance to win 100 000 yuan. The large reward is attracting a large number of lottery consumers. However, you only have a one-millionth chance to win the lottery. Please find the expected value of a lottery ticket.

Now we design a "fair lottery": You pay five yuan for a ticket, which contains one of two numbers — 1 and 2. With a ticket you are allowed to take a ball from a drawer, where there are 10 balls numbered 1 and 10 balls numbered 2, respectively. If the taken ball has the same number as that on your ticket, then you get 10 yuan; otherwise, you get nothing. Please find the expected value of a ticket of this kind. Furthermore, suppose the balls contained in the drawer still have two numbers, but we do not know how many ball there are in the two categories, respectively. Then what is the expected value of a ticket at this time?

Solution:

$$10^5 \times \frac{1}{10^6} = 0.1.$$

Therefore, the expected value of a lottery ticket in the first case is only 0.1 yuan. Since the price of a ticket is 1 yuan, so you lose averagely

$$1 - 0.1 = 0.9$$

yuan for each ticket you buy.

Some people think the more lottery tickets you buy, the larger the chance that you will make a fortune with it. As a matter of fact, however, the more tickets you buy, in average the more money you will lose. You need to buy 1 million tickets to ensure the 100 000 yuan reward. In the meantime, your loss is

$$10^6 - 10^5 = 9 \times 10^5.$$

You will lose 900 000 yuan.

Indeed, there is someone who buys a lottery ticket and earns the large reward. But it is an event with very small probability (only one to one-millionth), and his reward is actually provided by other lottery consumers. Those people and agencies selling lottery tickets are earning big money for nothing, a business more profitable even than those making money by small capital.

As for the case of fair lottery, the expected value of a ticket is

$$10 \times \left(\frac{1}{2} \times \frac{10}{10+10} + \frac{1}{2} \times \frac{10}{10+10}\right) = 10 \times \frac{1}{2} = 5(\text{yuan}).$$

Since the price for a ticket is also 5 yuan, so it is really a fair lottery. When the number of balls is unknown, we define the ratio of the number of number 1 balls to that of number 2 balls as p to q ($p + q = 1$). Then the expected value of a ticket is

$$10 \times \left(\frac{1}{2} \times p + \frac{1}{2} \times q\right) = 10 \times \frac{1}{2} = 5.$$

It is still 5 yuan.

But people or agencies selling this fair lottery tickets will make no money, and lottery consumers are reluctant to buy it as they think the reward is too little (10 yuan rather than 100 000 yuan) and the price is too high (5 yuan rather than 1 yuan). So such a fair lottery is not practicable. So many people prefer to buy a lot of lottery tickets with a very small chance to win a 100 100 yuan reward. It is an interesting psychological study topic.

57. Not to Indulge in Gambling

A gambler is allowed to select a number from 1, 2, 3, 4, 5, 6, and stake a yuan on it. Now, rolling three dice, if the selected number appears once, twice or three times, then the gambler will get money one, two or three times of the stake, respectively, and the original stake will also be returned to him. However, if the selected number does not appear, he will lose the stake.

Please find the expected value that the gambler will get.

Solution: We may assume that the number selected by the gambler is 1.

The probability that there is exactly one 1-spot in three dice is

$$3 \times \frac{1}{6} \times \frac{5}{6} \times \frac{5}{6} = \frac{3 \times 25}{216}.$$

The probability that there are exactly two 1-spot in three dice is

$$3 \times \frac{1}{6} \times \frac{1}{6} \times \frac{5}{6} = \frac{3 \times 5}{216}.$$

The probability that there are exactly three 1-spot in three dice is

$$\frac{1}{6} \times \frac{1}{6} \times \frac{1}{6} = \frac{1}{216}.$$

The probability that there is no 1-spot in three dice is

$$\frac{5}{6} \times \frac{5}{6} \times \frac{5}{6} = \frac{125}{216}.$$

Then we get the following table:

Money earned	a	$2a$	$3a$	$-a$
Probability	$\frac{3 \times 25}{216}$	$\frac{3 \times 5}{216}$	$\frac{1}{216}$	$\frac{125}{216}$

The expected value is then the weighted sum of a, $2a$, $3a$, $-a$, and that is

$$\frac{3 \times 25}{216} \times a + \frac{3 \times 5}{216} \times 2a + \frac{1}{216} \times 3a - \frac{125}{216} \times a = -\frac{17}{216}a.$$

So the gambler will lose $\frac{17}{216}a$ yuan each time. The percentage of it is

$$\frac{17}{216} \approx 7.87\%.$$

Usually the annual interest for money deposited in a bank is less than 2%. But the gambler will lose 7.87% for rolling dice each time, so his loss is really surprising. No wonder it is a common thing that many corrupt officials have easily lost millions of money in Macau casinos. As for salaried people, however, it is not worthy for them to give money earned by hard work to casinos. Not a few gamblers have ruined their family and themselves eventually.

58. Social Party

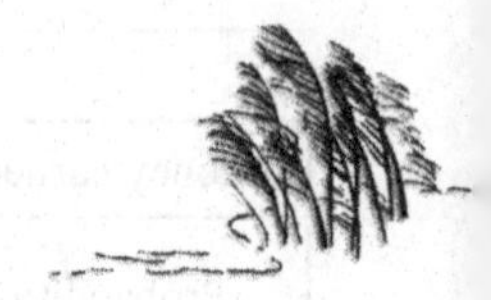

There are 8 girls and 7 boys in a social party, sitting around a round table. If all the girls sit together, there are then only two girls adjacent to boys. If girls and boys sit as alternately as possible, then there are 14 pairs of seats that are girl and boy adjacent. How many pairs of seats are there in average that are girl and boy adjacent?

Solution: In clockwise, there are two kinds of seat pairs that are girl and boy adjacent:

(G, B) and (B, G).

Therefore, the probability for a seat pair to be girl and boy adjacent is

$$\frac{8}{15} \times \frac{7}{14} + \frac{7}{15} \times \frac{8}{14} = \frac{8}{15}.$$

And the mathematical expectation for a seat pair to be girl and boy adjacent is

$$\frac{8}{15} \times 1 + \frac{7}{15} \times 0 = \frac{8}{15}.$$

As there are 15 seat pairs formed by 15 seats around the round table, so the average number of seat pairs that are girl and boy adjacent is

$$15 \times \frac{8}{15} = 8.$$

In general, if there are b boys and g girls sitting around a round table, then in average there are

$$(b+g) \times \left(\frac{b}{b+g} \times \frac{g}{b+g-1} + \frac{g}{b+g} \times \frac{b}{b+g-1}\right) = \frac{2bg}{b+g-1}$$

seat pairs that are girl and boy adjacent.

59. Success by Trying Once or More Times

Young man Wang takes a dice and wants to roll out the largest spot number — the 6-spot. But the result is not so. (We know that the probability to roll out the 6-spot with a dice is $\frac{1}{6}$, and that to get a contrary result is $\frac{5}{6}$.)

The famous ancient Chinese poet Lu You once said: "There has never been a success by trying once." So it is hardly possible to roll out the 6-spot by a dice only once.

However, the famous contemporary Chinese scholar Hu Shi said: "What Lu You said is not necessarily true", "All the successes from the ancient times to the present are achieved by trying." Then as long as you keep rolling dice, the 6-spot will appear in the end.

The famous Chinese mathematician Hua Luo-geng then commented: "'Try once' and 'trying more times' are basically two different concepts."

Putting aside the opinions of these three Chinese famous persons temporarily, we ask: How many times in average are needed for Wang to roll out the 6-spot?

Solution: Sometimes you get the 6-spot by rolling dice only once; sometimes you roll dice many times but the 6-spot fails to appear. So we cannot ask: "How many times of rolling dice are need to get the 6-spot?" But ask: "How many times of rolling dice in average are needed to get the 6-spot?"

The probability of getting the 6-spot by rolling dice only once is $\frac{1}{6}$.

The probability of getting the 6-spot by necessarily rolling dice twice is $\frac{5}{6} \times \frac{1}{6}$.

The probability of getting the 6-spot by necessarily rolling dice 3 times is $\left(\frac{5}{6}\right)^2 \times \frac{1}{6}$.

...

The probability of getting the 6-spot by necessarily rolling dice n times is $\left(\frac{5}{6}\right)^{n-1} \times \frac{1}{6}$.

Therefore, the average times of rolling dice are

$$M = 1 \times \frac{1}{6} + 2 \times \frac{5}{6} \times \frac{1}{6} + \cdots + n \times \left(\frac{5}{6}\right)^{n-1} \times \frac{1}{6} + \cdots. \quad (1)$$

By Eq. (1) $\times \frac{5}{6}$, we get

$$\frac{5}{6}M = 1 \times \frac{5}{6} \times \frac{1}{6} + \cdots + (n-1) \times \left(\frac{5}{6}\right)^{n-1} \times \frac{1}{6} + \cdots. \quad (2)$$

By (1) − (2), we get

$$\frac{1}{6}M = 1 \times \frac{1}{6} + \frac{5}{6} \times \frac{1}{6} + \cdots + \left(\frac{5}{6}\right)^{n-1} \times \frac{1}{6} + \cdots$$
$$= \frac{\frac{1}{6}}{1 - \frac{5}{6}} = 1.$$

So we have $M = 6$.

The probability $\frac{1}{6}$ can be replaced generally by p, and then the corresponding average value is $M = \frac{1}{p}$. For example, the average times of tossing a coin to get the head faced up is 2.

60. The 108 Heroes

A cigarette factory starts a promotion activity: Each pack of cigarettes will contain a picture of one of the 108 heroes from the Chinese traditional novel *The Water Margin*. If a consumer is able to collect all the 108 heroes by buying enough cigarettes produced by this factory, he will be given a reward. Please find the average number of packs of cigarettes the consumer should buy in order to win the reward.

Solution: After buying the first pack of cigarettes, he got a picture of heroes; after buying the second pack, he may get either a new picture or the same one as the first, and the probability to get a new picture is $\frac{107}{108}$. Therefore, by the reason showed in the previous chapter, he should buy $\frac{108}{107}$ packs to get the second picture. By the same reasoning, he should buy $\frac{108}{106}$ packs to get the third picture. By using the same reasoning continuously, we know that, in order to get all the 108 pictures, he should buy

$$1+\frac{108}{107}+\frac{108}{106}+\cdots+\frac{108}{1}=108\times\left(\frac{1}{1}+\frac{1}{2}+\cdots+\frac{1}{108}\right)$$

packs of cigarettes. It is not difficult to get, by a calculator, the value of the expression above as about 569.

We all know that smoking is harmful to health. It is unworthy to earn a reward at the cost of health. Besides, the factory sometimes plays tricks in the promotion: Cigarette packs in some region may contain less number of pictures of Song Jiang than normal, while in other regions may contain less Lu Jun-yi; or the picture of Wu Song is printed deliberately less than normal. Consequently it will be even harder to collect all the 108 heroes.

The number

$$\frac{1}{1}+\frac{1}{2}+\cdots+\frac{1}{108}$$

in the expression above is the sum of reciprocals of the first 108 natural numbers. The sum of reciprocals of the natural numbers

$$\frac{1}{1}+\frac{1}{2}+\cdots+\frac{1}{n}+\cdots$$

is called the harmonic series. It is a diverge series, i. e. it will tend to infinite as $n \to +\infty$. (An infinite series may not tend to infinite. For example,

$$1+\frac{1}{2}+\frac{1}{2^2}+\frac{1}{2^3}+\cdots=2$$

and

$$\frac{1}{1\times 2}+\frac{1}{2\times 3}+\frac{1}{3\times 4}+\cdots=1.$$

A series having finite sum like these is called a converge series.)

The great mathematician Leonhard Euler (1707 – 1783) has studied the harmonic series, giving the sum of its first n terms as

$$\frac{1}{1}+\frac{1}{2}+\cdots+\frac{1}{n}=\log_e n+0.577\,21+\varepsilon_n,$$

where 0. 577 21... has been called the Euler constant later and e = 2. 718 28... is the base of the natural logarithm (e is also the first letter of Euler); ε_n is very small and tends to zero as n increases.

61. Who is Sick?

An addicted gambler has a psychologist friend. The latter hopes to help the former in giving up gambling, and then offers a bet to him: Let the gambler play a gamble for 36 rounds. If he cannot win even once, then he should pay the psychologist 100 yuan and will gamble no more; otherwise, the psychologist will pay him 100 yuan. It is known that the bet for each round is 1 yuan — If the gambler wins, he can get 30 yuan as well as back the bet paid; otherwise, he will lose the bet. And the probability for him to win in a round is $\frac{1}{40}$. Please find the expected money the gambler will earn after the 36 rounds of gambling.

Solution: The mathematical expectation of each round is

$$30 \times \frac{1}{40} + (-1) \times \left(1 - \frac{1}{40}\right) = -\frac{9}{40} \text{ (yuan)}.$$

The probability for the gambler not to win even once is

$$\left(1 - \frac{1}{40}\right)^{36} = 0.401\,94\ldots,$$

and that to win at least one round is

$$1 - 0.401\,94\ldots = 0.598\,05\ldots.$$

Therefore, the expected money the gambler will earn is

$$100 \times 0.598\,05\ldots - 100 \times 0.401\,94\ldots + 36 \times \left(-\frac{9}{40}\right)$$
$$= 11.511\ldots \text{ (yuan)}.$$

The probability for the gambler to win at least one round in the gambling is more than one half, and he is expected to win more than

11 yuan. So this stop-gambling-by-gambling method is a bad way to help a gambler in giving up gambling. We cannot help to suspect that the psychologist is psychologically sick himself and need to see another psychologist.

62. A Fallen and Broken Rod

An evenly made glass rod with length l has fallen and broken into two fragments: One longer and the other shorter. We ask: How long in average is the longer fragment, as well as that of the shorter one?

Solution: Let the length of the shorter fragment be x, which could be any number in interval $\left[0, \frac{l}{2}\right]$. As there are infinite numbers in $\left[0, \frac{l}{2}\right]$, how can we find the average of them?

One way is by integral. We find the value of $\int_0^{\frac{l}{2}} x\,\mathrm{d}x$ first, and then divide it by $\frac{l}{2}$, i.e.

$$\frac{2}{l}\int_0^{\frac{l}{2}} x\,\mathrm{d}x = \frac{2}{l} \cdot \frac{1}{2}\left(\frac{l}{2}\right)^2 = \frac{l}{4}.$$

The other way is with the help of geometry. The numbers in $\left[0, \frac{l}{2}\right]$ may be regarded as points in it, so the average of these numbers is equivalently the center of the points, which is of course $\frac{l}{4}$.

The average length of the longer fragment can be found in the same way too. However, a simpler way is to calculate the length of the rod minus that in average of the shorter fragment, and that is

$$l - \frac{l}{4} = \frac{3}{4}l.$$

63. Broken into Three Fragments

Suppose the glass rod discussed in the last chapter has been broken into three fragments. How long averagely are the shortest, the longest, and the middle length fragments, respectively?

Solution: For simplicity, we may assume that the length of the rod is 1.

Suppose the lengths of three fragments are x, y, z, respectively, where x is the smallest. Then we have

$$x + y + z = 1, \tag{1}$$

$$x \leqslant y, \tag{2}$$

$$x \leqslant z. \tag{3}$$

Therefore,

$$z = 1 - x - y. \tag{4}$$

Substituting (4) into (3), we get

$$2x \leqslant 1 - y. \tag{5}$$

Since x and y are between 0 and 1, points (x, y) satisfying (2) and (5) are then located in the shade region in Fig. 1. The average value of these points is the coordinate of the center (denoted as G) of the region ($\triangle OCE$). Therefore, we are going to find the x-coordinate of G.

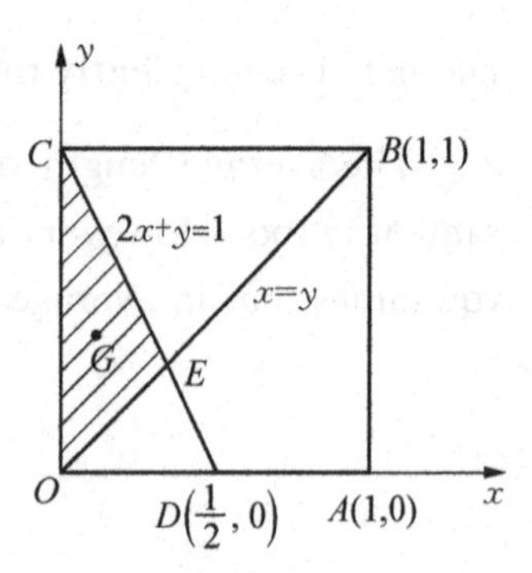

Fig. 1

It is not difficult to find that the x-coordinate of G is $\frac{1}{3}$ of E, which is in turn

$\frac{2}{3}$ of D $\left(\frac{CE}{CD}=\frac{2}{3}\right)$. Then the x-coordinate of E is $\frac{2}{3}\times\frac{1}{2}=\frac{1}{3}$, and that of G is

$$\frac{1}{3}\times\frac{2}{3}\times\frac{1}{2}=\frac{1}{9}. \tag{6}$$

So the average length of the shortest fragment is $\frac{1}{9}$.

In a similar way, suppose the lengths of three fragments are x, y, z, respectively, where x is the largest. Then (1) holds, and

$$x \geqslant y, \tag{7}$$

$$x \geqslant z. \tag{8}$$

Substituting (4) into (8), we get

$$2x \geqslant 1-y. \tag{9}$$

Besides, note that $z \geqslant 0$. Then from (4) we get

$$x+y \leqslant 1. \tag{10}$$

Points (x, y) satisfying (7), (9) and (10) are located in the shaded region in Fig. 2, and the average value of x is then the x-coordinate of the center G of it. This region consists of two triangles $\triangle DEF$ and $\triangle ADF$, and their area ratio is $1:3$ $\left(\frac{EF}{AF}=\frac{EF}{OF}=\frac{1}{3}\right)$. Then the x-coordinate of the center G_1 of $\triangle DEF$ is

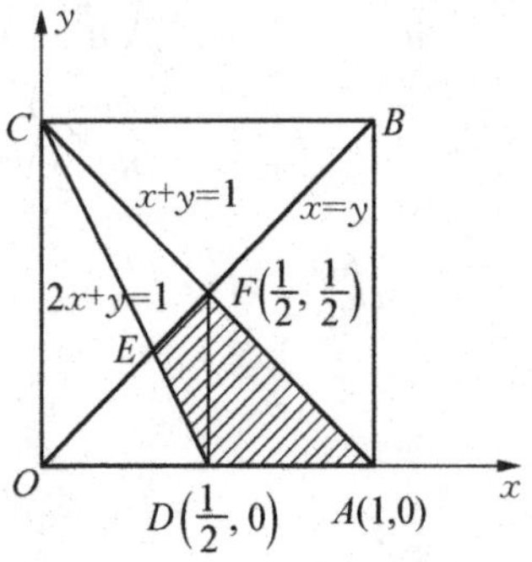

Fig. 2

$$\frac{1}{3}(x_D+x_E+x_F)=\frac{1}{3}\left(\frac{1}{2}+\frac{1}{3}+\frac{1}{2}\right)=\frac{4}{9},$$

and the x-coordinate of the center G_2 of $\triangle ADF$ is

$$\frac{1}{3}(x_A+x_D+x_F)=\frac{1}{3}\left(1+\frac{1}{2}+\frac{1}{2}\right)=\frac{2}{3}.$$

Since G divides G_1G_2 into $3:1$, we get the x-coordinate of G as

$$\frac{1\times\frac{4}{9}+3\times\frac{2}{3}}{1+3}=\frac{11}{18}.$$

Therefore, the average length of the longest fragment is $\frac{11}{18}$.

The average length of the second longest fragment is then

$$1-\frac{1}{9}-\frac{11}{18}=\frac{5}{18}.$$

If the length of the rod is l, then the average length of the three fragments is

$$\frac{1}{9}l,\ \frac{11}{18}l,\ \frac{5}{18}l,$$

respectively.

More generally, if a rod with length l is broken into n fragments, then the average length of each fragment, from the shortest to the longest, is

$$\frac{1}{n}\times\frac{1}{n},\ \frac{1}{n}\times\left(\frac{1}{n}+\frac{1}{n-1}\right),\ \frac{1}{n}\times\left(\frac{1}{n}+\frac{1}{n-1}+\frac{1}{n-2}\right),\ \ldots,$$
$$\frac{1}{n}\times\left(\frac{1}{n}+\frac{1}{n-1}+\cdots+\frac{1}{2}+1\right).$$

64. The Number of Cross Points

Suppose the side of each square on a squared paper is 1. Now drop a needle with length l on the paper. Please find the average number of points at which the needle and lines meet.

Solution: If $l < 1$, then by using the result of the Buffon's needle problem in Chapter 44, we have that the probability the needle lies across the horizontal lines of the paper is $\frac{2l}{\pi}$, and the number of cross points is 1, because it is impossible for the needle to cross two horizontal lines. Therefore, the average number of cross points produced by the needle on the horizontal lines is

$$\frac{2l}{\pi} \times 1 + \left(1 - \frac{2l}{\pi}\right) \times 0 = \frac{2l}{\pi}.$$

By the same reason, the average number of cross points produced by the needle on the vertical lines is also $\frac{2l}{\pi}$.

Consequently, the average number of cross points produced by the needle is

$$\frac{2l}{\pi} + \frac{2l}{\pi} = \frac{4l}{\pi}. \tag{1}$$

If $l \geqslant 1$, we divide the needle into several segments, with each length less than 1. Suppose the length of each segment is $l_1, l_2, \ldots, l_n$, respectively. Then the average number of cross points produced by each segment is

$$\frac{4l_k}{\pi} \ (k = 1, 2, \ldots, n).$$

Therefore, the average number of cross points produced by the entire needle is

$$\sum_{k=1}^{n} \frac{4l_k}{\pi} = \frac{4l}{\pi}, \tag{2}$$

which means expression (1) still holds.

65. Throw a Wire Ring

There are equal-distance parallel lines on a flat plane. Suppose the distance is a. Now throw a wire ring with circumference $2l$ ($2l < \pi a$) onto the plane. Please find the probability for the ring lying across lines.

Remark: This is a generalization of the Buffon's needle problem.

Solution: We divide the wire ring into several segments, with each length

$$l_1, l_2, \ldots, l_n \ (l_1 + l_2 + \cdots + l_n = 2l),$$

being less than a, respectively. When the segments are small enough, they can be regarded as straight ones. Then according to the solution of the Buffon's needle problem, the probability that the kth segment lies across a line is

$$\frac{2l_k}{\pi a} \ (k = 1, 2, \ldots, n). \tag{1}$$

In a similar way as shown in the last chapter, the average number of the cross points produced by the kth segment is $\frac{2l_k}{\pi a}$, and that produced by the entire wire ring is $\frac{4l}{\pi a}$.

Since the diameter of the wire ring is $\frac{2l}{\pi} < a$, it lies across one line at most. Therefore, the number of cross points is either 0 or 2. (In the case that a line is tangent to the ring, the two cross points are regarded as coincided.)

Denote the probability that the ring lies across the lines by p.

Then the average number of cross points produced by the ring is

$$2 \times p + 0 \times (1 - p) = 2p.$$

Comparing it with the expression above, we get that

$$2p = \frac{4l}{\pi a}. \tag{2}$$

Therefore,

$$p = \frac{2l}{\pi a}. \tag{3}$$

Please note: If we add each probability in (1) (i.e. the probability that a specific segment lies across a line) directly, we will get $\frac{4l}{\pi a}$, which doubles the value of (3). It is because the events that segments lie across the parallel lines are not independent with one another, therefore the sum of the probabilities of these events will be greater than that the entire wire ring lies across the lines.

On the other hand, the average values (i. e. the mathematical expectations) are addible (no matter if the corresponding random variables are independent with each other or not). So we are able to use the roundabout way above to obtain the required probability correctly.

The ring in this problem can be generalized to convexities (e. g. triangles, convex quadrilaterals, ellipses, etc.), while the corresponding condition should be changed from $2l < \pi a$ to "the width of the convexity $< a$ ".

Exercises

1. Take out 3 balls from a bag containing 4 white and 5 black ones. Please find the probability that all the three balls are black.

2. Roll 2 dice once. What is the probability that the 2-spot appears at least once?

3. Roll 3 dice once. What is the probability that the sum of the dice is 14?

4. Given a glass cup made by a factory, the probability that it is broken on its first drop is $\frac{1}{2}$. If it has not been broken, then the probability that it is broken on the second drop is $\frac{7}{10}$. If it still not been broken, then the probability that it is broken in the third drop is $\frac{9}{10}$. Please find the probability that the cup has not been broken after three drops.

5. Shooting a plane with a rifle, the probability of hit is 0.004. Now shooting a plane with 250 rifles, what is the probability of hit at this time?

6. The rule of a pass-through-gate game is: When reaching the nth gate, you need rolling a dice n times. If the sum of the n spot numbers is greater than 2^n, then you are allowed to pass through the gate; otherwise, you are not allowed. We ask:

(1) what is the highest level gate that you have a chance to pass through?

(2) what is the probability that you pass through the first three gates?

7. Roll 4 dice once. Please find the probability that the sum of the spot numbers is 10 or 20.

8. In pot A there are 2 red and 1 black balls, and in pot B there are 101 red and 100 black balls. Now take out a ball from a pot and look at it. If the ball is black, it will be put back into the pot; if it is red, then it will not be put back. Then take out another ball from the same pot and look it. Finally you should guess whether the pot from which balls are taken is A or B. What is the probability that you guess correctly?

9. Rolling 4, 3 and 2 dice once, respectively, and each sum of the spot numbers is equally 6. Please prove that the ratio of the probabilities of these three events is $1:6:18$.

10. There are 3 sets of books, each containing 3, 4 and 1 ones, respectively. Now put these 8 books in a row on the bookshelf. Please find the probability that each set of books are put together, respectively.

11. Roll 6 dice once. Please compare the probability that all the spot numbers are the same with that they are different from each other.

12. Take out 3 balls from a bag that contains 5 white and 8 black ones, then put them back, and finally take out 3 balls from the bag. Please find the probability that the first 3 balls are white and the last 3 ones are also white.

13. There are 1 gold and 3 copper coins in the first bag, 2 gold and 4 copper coins in the second bag, and 3 gold and 1 copper coins in the third bag, respectively. Please find the probability that a coin taken out randomly from one of the three bags is gold.

14. The first 20 positive integers have been written on 20 paper slips, respectively. Now take a slip randomly from them. Please find the probability that the number on it is divisible by either 3 or 5.

15. There are 5 white and 7 black balls in a bag. Take out 2 balls from the bag. What is the probability that they are one white and one black?

16. Multiply four natural numbers selected randomly. Please find the probability that the unit digit of the product is 1, 3, 7 or 9.

17. Roll two dice 25 times continuously. Please compare the probability that the "double six" (i. e. both spot numbers are 6) appears at least once with that it does not appear even once.

18. Please find and compare the probabilities of two events: (1) roll a dice 4 times and the 6-spot appears at least once; (2) roll two dice 24 times and the "double six" appears at least once.

19. In a Show Hand card game, a pack of 52 cards are dealt to the participants, so that each of them gets a set of 5 cards. Please find the probabilities of the following events, respectively.

(1) The set consists of A, K, Q, J and 10 in the same suit.

(2) The set is four of a kind.

(3) The set consists of three of a kind and one pair.

(4) The set is straight.

(5) The set is three of a kind.

(6) The set is two pairs.

(7) The set is one pair.

20. From the digits 1, 2, 3, 4 and 5 select 3 ones to form a three-place number. Please find the probability that it is an even number.

21. Using the principle of probability theory, prove the identical equation

$$1+\frac{A-a}{A-1}+\frac{(A-a)(A-a-1)}{(A-1)(A-2)}+\cdots+\frac{(A-a)\cdots 2\cdot 1}{(A-1)\cdots(a+1)a}=\frac{A}{a},$$

where A, a are positive integers, with $A>a$.

22. Take out a telephone number randomly from a telephone directory. Please find the probability that its last four digits are different from each other.

23. There are 10 football players numbered from 1 to 10. A journalist has interviews with any three of them. Please find the probabilities that (1) the smallest number for the three is 5; and (2) the largest number for the three is 5.

24. There are 3 000 products, among which 2 200 ones are qualified and the other 800 unqualified. Now take randomly 400 ones from them. Please find the probabilities that in these 400 products: (1) there are exactly 180 unqualified; and (2) there are at least 2 unqualified.

25. A company has 17 boxes of products, among which 10 boxes are of the first class, 4 ones of the second class, and 3 ones the third class. Now a client needs 4 boxes of the first class, 3 ones of the second class, and 2 ones the third class. As the labels attached to each box are all gone, the delivery man has to deliver the products randomly. Please find the probability that the client gets just the products he wants.

26. Take randomly 4 shoes from 5 pairs of different shoes. Please find the probability that there are at least one pair of shoes in them.

27. There are a batch of 100 products, among which 5 ones are unqualified while all the others are qualified. Now take 50 products from the batch. Please find the probabilities that among them: (1) these is no any unqualified product, and (2) there is at least one unqualified product.

28. There are a batch of pan products, among which 16 ones are of the first class and the other 4 are of the second class. Now take any 3 pans from them. Please find the probabilities that among the three:

(1) there is at least one pan that is of the first class,

(2) there is at least one pan that is of the second class,

(3) all are of the first class, and

(4) all are of the second class.

29. Among 10000 lottery tickets there is one being the first prize, 10 the second prize, 50 the third prize, and 100 the fourth prize. The rewards of the first, second, third, and fourth prizes are 500 yuan, 100 yuan, 20 yuan, and 5 yuan, respectively. Someone has bought a lottery ticket. Please find the probability that he gets a reward not less than 20 yuan.

30. There are 20 cars, including 5 passenger ones, made by the first car factory; there are 8 cars, including 4 passenger ones, by the second car factory; and there are 12 cars, all being passenger ones, by the third car factory. Suppose the probability for each car running on the highway is the same. Now it is found that a car of the first factory is running on the highway. What is the probability that it is a passenger car?

31. There are 2 white and 3 black balls in a pot. Now take out two balls from the pot one by one. Please find the probability that they are both white. If the first ball taken out has been put back before taking out the second one, then what is the probability that both are white balls?

32. Take randomly a number x from the integers between 1 and 100. Please find the probability that $x+\frac{100}{x}>50$.

33. There are 250 trees in a residential district, among which 135 ones are larches, 68 are spruces, and 47 white birches. Select randomly a tree from them. Please find the probability that it is a spruce, and that it is not a white birch.

34. There are 800 seeds from batch A, whose sprouting percentage is 80%; and there are 1 000 seeds from batch B, whose sprouting percentage is 70%. Now mix the seeds of two batches and then take out one from them. Please find the probability that it will sprout.

35. There are 100 seedlings infected with disease germs in a nursery garden, and 20% of them is dead. Now take randomly 4 seedlings from all the infected ones. Please find

(1) the probability that all four are dead, and

(2) the probability that there are exactly 2 dead seedlings among the four.

36. Suppose the probability for the sample yarn of the first kind to pass the standard load test is 0.84, and that for the second kind is 0.78. Now take one sample yarn from each kind, respectively. Please find the probability that both yarns pass the standard load test.

37. A plane consists of three parts. It will be shot down, if the first part is shot by one bullet, or the second part shot by two bullets, or the third part shot by three bullets. And the probability that a part is shot is in direct proportion to the area of it. Suppose the ratio of the areas of three parts is 1 : 2 : 7. Please find

(1) the probability that the plane is shot down by two bullets, and

(2) the probability that it is shot down by three bullets.

38. Two planes bomb a target in turn, each time dropping one bomb. Suppose each plane carries three bombs, and the probabilities for them to hit the target are 0.3 and 0.4 respectively. Please find the probability that they hit the target before they have dropped all the bombs.

39. Whcn a shell explodes at a distance of R from a plane, the condition to destroy the plane is if it hits two engines or the cockpit. It is known that the probability to hit each engine is 0.2, and that to hit the cockpit is 0.3. Please find the probability that the shell destroys the plane.

40. Shoot a plane three times. The probability of being hit the first time is 0.4, that in the second time is 0.5, and that in the third time is 0.7. The probability that the plane is shot down by one hit is 0.2, that by two hits is 0.6, and that by three is 1. Please find

(1) the probability that the plane is shot down by three hits, and

(2) the probability that the plane is hit once when it was shot down.

41. The statistics of a football season shows that: the probabilities of shooting at 25 m, 20 m, and 15 m from the goal are 0.1, 0.7, and 0.2, with the probabilities of scoring the goal being 0.05, 0.1, and 0.2, respectively. It is known that a goal is scored. Please find the probability that it is shot at 20 m from the goal.

42. Two shooters shoot the same target independently. The hit rate of the first shooter is 0.8, and that of the second shooter is 0.4. It is found that the target is hit only once after the two finish their shooting. Please find the probability that it is shot by the first shooter.

43. Shoot a plane four times. The hit rate of each shooting is 0.3. The probability the plane is shot by one hit is 0.6, and that by two hits is 1. Please find the probability that the plane is shot down.

44. The hit rate of each shooting is 0.2. A target will burn if it is hit at least twice. Now shoot the target 8 times. Please find the probability that it burns.

45. Roll 2 dice once and the sum of spot numbers is 7. Please find the probability that one of the spot numbers is 1.

46. The 11 letters that form the word "probability" have been written on 11 cards, respectively. Now randomly take 7 cards one by one from them. Please find the probability that the sequence of them forms the word "ability".

47. Put three balls into four drawers randomly. Please find the probabilities that the largest number of balls in a drawer is 1, 2, and 3, respectively.

48. A total of 50 staff members are sent to 10 sections. Among them there are three members who are not able to work together in any section. Please find the probability that they are sent to the same section.

49. A and B roll a dice in turn, and A does it first. If A gets 6-spot, he then win the game; otherwise, B rolls the dice. If B gets 6- or 5-spot, he wins the game; otherwise, A rolls the dice. If A gets 6- or 5- or 4-spot, he wins, otherwise, and so on. The game goes on in this way. Please find the probabilities for each to win.

50. A number consists of 7 digits whose sum is 59. Please find the probability that the number is divisible by 11.

51. There are 10 tickets. Among them 5 ones have been marked "0", while the other 5 ones marked "1", "2", "3", "4" and "5", respectively. Now take three tickets one by one with (1) all of them not being put back, and (2) the previous ticket being put back before taking next one. Please find the related probabilities that the sum of digits on the three tickets is 10 in two cases, respectively.

52. Take n integers randomly, and multiply them. Please prove:

the probability for the digit in the ones place of the product being 1, 3, 7 or 9 is $\frac{2^n}{5^n}$, that being 2, 4, 6 or 8 is $\frac{4^n - 2^n}{5^n}$, that being 5 is $\frac{5^n - 4^n}{10^n}$, and that being 0 is $\frac{10^n - 8^n - 5^n + 4}{10^n}$, respectively.

53. Ten things are allocated to three persons A, B and C. Please find the probability that A gets more than 5 things.

54. It is known that two events A and B satisfy: $P(A) = \frac{1}{6}$, $P(B \mid A) = \frac{1}{2}$, and $P(A \mid B) = \frac{1}{3}$. Please find $P(A \cup B)$.

55. It is known that $P(A) = 0.7$, $P(\bar{B}) = 0.6$ and $P(A\bar{B}) = 0.5$. Find $P(\bar{B} \mid \bar{A} \cup B)$.

56. As shown in the below below, the close probabilities of switches 1, 2 and 3 are p equally, and they are independent with each other. Please find the probability that nodes A and B are connected.

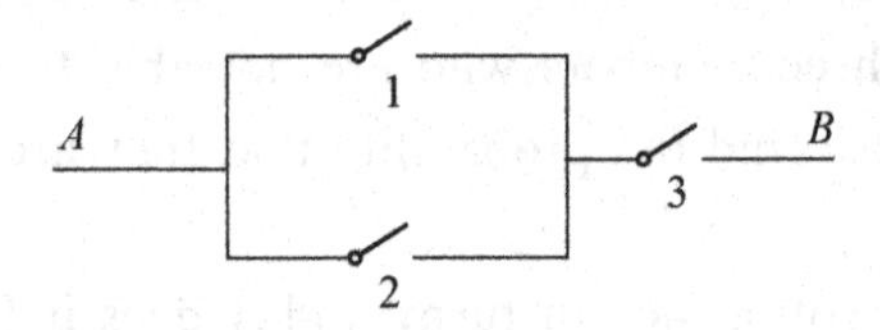

57. The condition is the same as that in the previous question. Please find the probability that nodes A and B as shown in the figure below are connected.

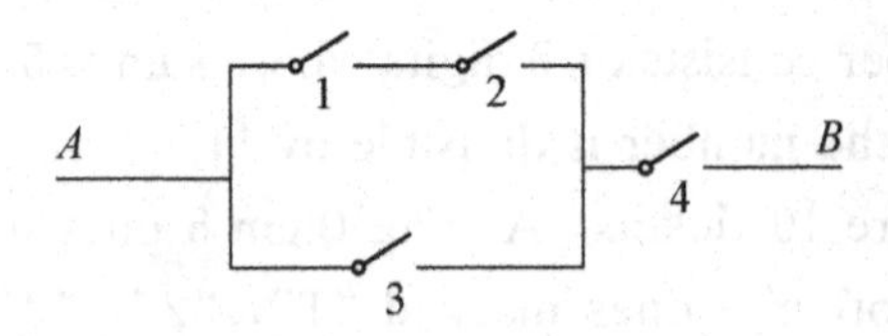

58. The condition is the same as that in Exercise 56. Please find the probability that nodes A and B as shown in the figure on the next page are connected.

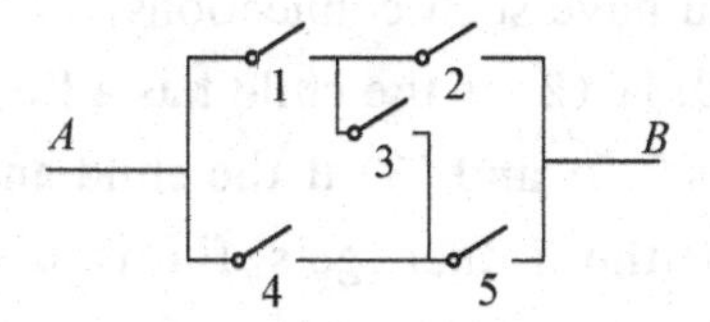

59. There are 9 white and 1 red balls in a bag. Ten persons each takes out one ball from the bag in turn, and no ball will be put back. Please find the probabilities for each person to get the red ball, respectively.

60. There are a white and b red balls in bag A, and there are c white and d red balls in bag B. Now take out a ball from bag A and put it into bag B. Then take out a ball from bag B. What is the probability that the ball is white?

61. While dialing, a man cannot remember the last digit of the telephone number. He has to try it. What is the probability he dials the phone successfully by just four tries? And what is the probability that he succeeds with at most four tries?

62. Three persons each tries to break a code independently. It is known that the probabilities of success for them are $\frac{1}{3}$, $\frac{1}{4}$ and $\frac{1}{5}$, respectively. Please find the probability that at least one of them breaks the code.

63. In a bag, there are 10 real coins, as well as 5 counterfeit coins with both sides being head. Take out a coin randomly from the bag, toss it three times, and it lies on the ground with the head side facing up each time. Please find the probability that the coin is a real one.

64. It is known that a batch of 10 products contain 2 unqualified ones. Now take two products one by one from them, with none being put back. Please find the probabilities of the following events:

(1) both are qualified,

(2) one is qualified and the other unqualified,

(3) the second one is unqualified.

65. According to the past data, the events concerning a three-person family and flu have such connections: (1) the probability that the child gets flu is 0.4; (2) if the child has a flu, the probability that the mother gets flu is 0.5; and (3) if the child and mother have a flu, the probability that the father gets flu is 0.4. Please find the probability that both the child and mother get flu but the father does not.

66. Among a group of people, 37.5% persons have blood type A, 20% have blood type B, 35% blood type O, and the others blood type AB. The blood donation match table is shown in the following, where symbol "√" means the related donator and recipient are matched (i.e. the former's blood can be transfused to the latter).

Blood recipient	Blood donor			
	Type A	Type B	Type AB	Type O
Type A	√	×	√	√
Type B	×	√	√	√
Type AB	√	√	√	√
Type O	×	×	×	√

From them, select a person as a blood donor, and another one as a blood recipient. Please find the probability that they are matched.

67. It is known that 5% of men and 0.25% of women are color-blind. Select a person from 100 men and 100 women. If he/she is color-blind, what is the probability that this is a man?

68. When we input one of three letters A, B and C, the probability that the output of the letter inputted is α, and that it is one of the other two letters is $\frac{1-\alpha}{2}$. Now we input one of three letter strings $AAAA$, $BBBB$ and $CCCC$, with probability of p_1, p_2 and p_3 ($p_1 + p_2 + p_3 = 1$), respectively, and get the output $ACBA$. Please find the probability that the input is $AAAA$ (assume that each input is independent).

69. Take any three points on a circle. Please find the probability that they constitute an acute triangle.

70. Take any three points on a circle, and cut it into three arcs. Please find the probability that the sum of any two arcs is larger than the other one.

71. Take any two points from a line segment of length a. Please find the probability that the distance between two points is larger than $b(b < a)$.

72. Take two points randomly from a line segment to cut it into three parts. Please find the probability that these three parts constitute a triangle.

73. Mark n points randomly on a rod to divide it into $n+1$ segments. Please find the probability that the length of each segment is not greater than $\frac{1}{n}$ of the rod.

74. Take two points randomly from a segment of length a to cut it into three parts. Please find the probability that the length of each part is not greater than constant b $\left(\frac{a}{3} < b < a\right)$.

75. Each of the two segments of length a have been cut off some part randomly. Please find the probabilities that the sum of two remaining segments is (1) greater than a, (2) less than a, and (3) equal to a, respectively.

76. Take three segments randomly. Please find the probability that they form a triangle.

77. Given $\odot O$, make three tangents to it. Then it can be either an inscribed or an escribed circle of the triangle formed by these three tangents. Please prove that the ratio of the probabilities of the two cases is 1 : 3.

78. On a line segment of length $a+b$, measure randomly two segments with lengths a and b respectively. Please prove that the probability for the common part of these two segments being not greater than c is $\frac{c^2}{ab}(c < b, a)$, and that for the shorter segment (of length b) lying completely in the longer one (of length a) is $\frac{a-b}{a}$.

79. On a line segment of length $a+b+c$, measure randomly two segments with lengths a and b in respect. Please prove: (1) the probability that two segments have no common part is $\frac{c^2}{(c+a)(c+b)}$, and (2) that they have a common part with a length not greater than d is $\frac{(c+d)^2}{(c+a)(c+b)}$ $(d < a, b)$.

80. There are a group of parallel lines with distance d on a plane. Now throw a triangle with sides a, b, c (all shorter than d) onto the plane. Please find the probability that the sides of the triangle lie across the parallel lines.

81. There are five 10-yuan notes and one 20-yuan note in the first bag, and six 10-yuan notes in the second bag. Now take two notes from the first bag and put them into the second one; then take two notes from the second bag and put them into the first one. Please find the mathematical expectations of the money in each bag.

82. Gamblers A and B roll a dice in turn. Whoever gets the 6-spot first will win the bet of 110 yuan. A has the first turn. Please find the values of money they are expected to win, respectively.

83. There are two 100-yuan and three 50-yuan notes in a bag. A man takes out two notes from it. Please find the value of money he is expected to get.

84. A man has one 100-yuan and four 10-yuan notes, and he takes out two notes randomly to give his grandson and granddaughter, respectively. Please find the value of money the grandson is expected to get.

85. A and B possess gambling techniques which are in equal power. They are having a series of gambling rounds: For each round the winner will get 1 point and there will be no draw; whoever wins s points first will win the gambling. The stake is 160 yuan. Suppose the gambling stops when A has $s-3$ points and B has $s-2$ points. How should the stake be allocated between them?

86. A has a bet with B : the stake is A's 5 yuan plus B's 2 yuan; whoever wins the bet will take all the 7 yuan. They each roll 2 dice simultaneously. If A gets 7 spots before B gets 4 spots, then A wins the bet; otherwise, B wins the bet. If A gets 7 and B gets 4 spots simultaneously, then the bet continues till someone wins. Please find the mathematical expectation of B.

87. There are one M-yuan note and several m-yuan notes in a bag. Take out one note from the bag, till you get the note with M-yuan. Please find the mathematical expectation of the money taken out.

88. Shoot at a target continuously, till the target is hit. The probability of hit for each shot is p. Please find the average number of shots.

89. Calculate $\sum_{k=0}^{n} k^2 C_n^k p^k q^{n-k}$ and $\sum_{k=0}^{n} (k - np)^2 C_n^k p^k q^{n-k}$, where $q = 1 - p$.

90. There are n students in a class. Suppose the number of days when it is exactly k students' birthday is x. Please find the mathematical expectation of x.

91. Referring to the exercise above, please find the mathematical expectation of the number of days when it is 2 or more students' birthday.

92. Shoot a target n times, and the probability of hit for each time is p. Prove that the hit number m satisfies

$$P\left(\left|\frac{m}{n} - p\right| > \varepsilon\right) < \frac{1}{4n\varepsilon^2},$$

where ε is any given positive number.

93. Calculate $\sum_{k=0}^{n} k^3 C_n^k p^k q^{n-k}$, where $q = 1 - p$.

94. Let X be a random variable. Please prove

$$E[(X - EX)^2] \leqslant E(X^2).$$

When will the equality hold?

95. Catch n fishes from a lake, mark them and release them back to the lake. Then catch r fishes again, and find out among them k ones are marked. Suppose there are N fishes in the lake. Given n, r, k, please find N so that the required probability is the maximum.

96. Three chess players A, B, C start a round competition. The first round is A versus B. Then the winner meets C in the next round. The new winner will contest with the player who lost in the previous round. The competition will continue in this way, till someone wins twice successively, who then is the champion.

(1) Suppose three players are at the same level of power. What are the probabilities for them to win the competition in respective?

(2) If A wins the first round, what are the probabilities for them to win the competition in respective?

97. On the perimeter of a given circle, select randomly six points A, B, C, D, E, F. Please find the probability that the sides of $\triangle ABC$ do not cross that of $\triangle DEF$.

98. Given the same three dice with n sides, on each of their corresponding sides, write the same integer. Roll them randomly. Please prove that the probability for the sum of the integers on the sides facing up divisible by 3 is not less than $\frac{1}{4}$.

99. Toss a coin, if the result is the head side facing up, we record one point; if it is the tail side facing up, then record two points. Toss and record in this way continuously. Please prove that the probability of getting exactly a sum of n points is $\frac{2+\left(-\frac{1}{2}\right)^n}{3}$.

100. There is a pack of N cards, including three aces. After shuffling randomly, take the cards one by one from the top, till you get the second ace. Please prove that the mathematical expectation of the number of cards taken is $\frac{N+1}{2}$.

101. There is more than one hundred balls with red or black color in a pot. If we take out two balls from the pot randomly, the probability that both are red is $\frac{1}{2}$. How many red balls are there in the pot?

102. 60 male and 100 female turtles are put into five jars in such a way that, when we take out one from the jars randomly, the probability to get a male turtle is the greatest. What is this greatest probability?

103. Prove $\sum_{k=0}^{\infty} \frac{1}{3}\left(\frac{2}{9}\right)^k \frac{C_{2k}^k}{k+1} = \frac{1}{2}$. More generally, let $0 \leqslant p \leqslant 1$, $q = 1 - p$. Prove

$$\sum_{k=0}^{\infty} (pq)^{k+1} \frac{C_{2k}^k}{k+1} = \begin{cases} p, & \text{when } p < \frac{1}{2}, \\ q, & \text{when } p \geqslant \frac{1}{2}. \end{cases}$$

104. A fishing ship fishes furtively on the water of a neighboring country. Each time it will gain m yuan. But if it is discovered by a patrol boat of the neighboring country, all the fish caught will be confiscated (i. e. it will gain 0 yuan). The probability for it to be caught is $\frac{1}{k}$ (k is a positive integer). In order to make the mathematical expectation of gaining the maximum, how many times should the ship fish there?

105. Tossing a coin, we get one point if it is the head side facing up, and two points if the tail side facing up. Please find the probability that we get exactly n points.

106. Suppose $\frac{n(n+1)}{2}$ different numbers are arranged into a triangle as follows.

$$\begin{array}{c} * \\ * \quad * \\ * \quad * \quad * \\ \cdots \\ * \quad * \cdots * \quad * \end{array}$$

Let M_k be the maximum number in the kth row (in order from top to bottom). Please find the probability that $M_1 < M_2 < \cdots < M_n$.

107. A particle moves between places A and B following the rule:

(1) When it is in place B, it will move to A in 1 second.

(2) When it is in place A, it will stay in A or move to B with probability $\frac{1}{2}$, respectively, in 1 second.

Now the particle is in place A. Please find the probability that it stays in A after 10 seconds.

Key to Exercises

1. $\frac{C_5^3}{C_9^3} = \frac{5}{42}$.

2. $1 - \frac{5}{6} \times \frac{5}{6} = \frac{11}{36}$.

3. When the sum is 14, the possible numbers of spots distributed on three dice may be (6, 6, 2), (6, 5, 3), (6, 4, 4) and (5, 5, 4). Therefore, there are $3 + 3! + 3 + 3 = 15$ possible results. The probability is then $\frac{15}{36} = \frac{5}{12}$.

4. $\left(1 - \frac{1}{2}\right)\left(1 - \frac{7}{10}\right)\left(1 - \frac{9}{10}\right) = \frac{3}{200} = 0.015$.

5. $1 - (1 - 0.004)^{250} = 1 - 0.996^{250} \approx 1 - 0.37 = 0.63$.

6. (1) Since $2^5 = 32 > 6 \times 5 = 30$, it is impossible for us to pass through the fifth level gate. Therefore, we are able to pass through at most the fourth level gate.

(2) In order to pass through the first level gate, the spot number should be greater than 2, whose probability is $\frac{4}{6}$.

In order to pass through the second gate, the sum of two spot numbers should be greater than 4. As there are 6 possible results with each sum not greater than 4 (i.e. (1, 1), (1, 2), (2, 1), (1, 3), (3, 1), (2, 2)), therefore the probability is $1 - \frac{6}{36} = \frac{30}{36}$.

To pass through the third gate, the sum of three spot numbers should be greater than 8. As there are 56 possible results with each sum not greater than 8, therefore the probability is $1 - \frac{56}{6^3} = \frac{160}{216}$.

So the required probability is

$$\frac{4}{6} \times \frac{30}{36} \times \frac{160}{216} = \frac{100}{243}.$$

7. Rolling 4 dice once, there are 35 possible results with each sum of the spot numbers being 20, and 80 possible results with each sum being 10. Therefore, the required probability is

$$\frac{35+80}{6^4} = \frac{115}{1\,296}.$$

8. In a similar way as shown in Chapter 16, we construct a table as follows:

Balls taken out	The probability they are from pot A		The probability they are from pot B	You should guess
Red, red	$\frac{1}{2} \times \frac{2}{3} \times \frac{1}{2}$	$>$	$\frac{1}{2} \times \frac{101}{201} \times \frac{100}{200}$	A
Red, black	$\frac{1}{2} \times \frac{2}{3} \times \frac{1}{2}$	$>$	$\frac{1}{2} \times \frac{101}{201} \times \frac{100}{200}$	A
Black, red	$\frac{1}{2} \times \frac{1}{3} \times \frac{2}{3}$	$<$	$\frac{1}{2} \times \frac{100}{201} \times \frac{101}{201}$	B
Black, black	$\frac{1}{2} \times \frac{1}{3} \times \frac{1}{3}$	$<$	$\frac{1}{2} \times \frac{100}{201} \times \frac{100}{201}$	B

The probability that you guess correctly is

$$\frac{1}{2} \times \left(\frac{1}{3} + \frac{1}{3} + \frac{100 \times 101}{201 \times 201} + \frac{100 \times 100}{201 \times 201}\right) = 0.582\,08\ldots.$$

9. Rolling 2 dice once, the spot number figures satisfying the sum 6 are (1, 5), (2, 4) and (3, 3), which correspond to 5 possible results. So the probability is $\frac{5}{6^2}$. Rolling 3 dice once, the spot number figures satisfying the sum 6 are (1, 1, 4), (1, 2, 3) and (2, 2, 2), corresponding to 10 possible results. So the probability is $\frac{10}{6^3}$. Rolling 4 dice once, the spot number figures satisfying the sum 6 are (1, 1, 1, 3) and (1, 1, 2, 2), corresponding to 10 possible results. So the probability is $\frac{10}{6^4}$.

Therefore, the ratio of three probabilities is

$$\frac{10}{6^4} : \frac{10}{6^3} : \frac{5}{6^2} = 1 : 6 : 18.$$

10. There are 8! permutations of 8 books in a row. On the other hand, there are 3! permutations of 3 sets, 3! permutations of 3 books in the first set, and 4! permutations of 4 books in the second set. Therefore, there are 3!3!4! permutations satisfying the condition. So the required probability is $\frac{3!3!4!}{8!} = \frac{3}{140}$.

11. The probability that the 6 spot numbers of the six dice are the same is $\frac{6}{6^6}$, while that they are different from each other is $\frac{6!}{6^6}$. So it is obvious that the latter is larger.

12. There are C_{13}^3 possible results to take out 3 balls from the bag, among which there are C_5^3 ones consisting of 3 white balls. Therefore, the required probability is

$$\left(\frac{C_5^3}{C_{13}^3}\right)^2 = \left(\frac{5}{143}\right)^2 = \frac{25}{20\,449}.$$

13. The chance for each bag to be selected is $\frac{1}{3}$; the probabilities to take out a gold coin from the first, second and third bags are $\frac{1}{4}$, $\frac{2}{6}$ and $\frac{3}{4}$, respectively. So the required probability is

$$\frac{1}{3} \times \left(\frac{1}{4} + \frac{2}{6} + \frac{3}{4}\right) = \frac{4}{9}.$$

14. There are $\left[\frac{20}{3}\right] = 6$ multiples of 3, $\left[\frac{20}{5}\right] = 4$ multiples of 5, and 1 multiple of 15, respectively. So the required probability is

$$\frac{6+4-1}{20}=\frac{9}{20}.$$

15. There are C_{12}^2 possible results to take out 2 balls from the bag, among which there are 5×7 ones consisting of one white and one black balls. So the required probability is

$$\frac{5\times 7}{C_{12}^2}=\frac{35}{66}.$$

16. The ones place of each of the four natural numbers must be 1, 3, 7 or 9, whose probability is $\frac{4}{10}$. Therefore, the required probability is

$$\left(\frac{4}{10}\right)^4=\frac{16}{625}.$$

17. The probability that the "double six" does not appear even once is $\left(1-\frac{1}{36}\right)^{25}=0.505\ldots>\frac{1}{2}$. Therefore, the latter probability is larger.

18. The probability to roll a dice once and the 6-spot appears at least once is

$$1-\left(\frac{5}{6}\right)^4=0.51\ldots.$$

The probability to roll two dice 24 times and the "double six" appears at least once is

$$1-\left(\frac{35}{36}\right)^{24}=0.49\ldots.$$

So the former is larger.

19. Let $p=\frac{1}{C_{52}^5}$. Then the required probabilities are, respectively,

(1) $4p=\frac{1}{649\,740}$;

(2) $13 \times (52 - 4)p = 624p = \frac{1}{4\,165}$;

(3) $13 \times C_4^3 \times 12 \times C_4^2 p = 3\,744p = \frac{6}{4\,165}$;

(4) $9 \times 4^5 p = 9\,216p = \frac{192}{13 \times 4\,165}$;

(5) $13 \times C_4^3 \times C_{12}^2 \times 4^2 p = 54\,912p = \frac{88}{4\,165}$;

(6) $C_{13}^2 \times (C_4^2)^2 \times (52 - 2 \times 4)p = 123\,552p = \frac{198}{4\,165}$;

(7) $13 \times C_4^2 \times C_{12}^3 \times 4^3 p = 1\,098\,240p = \frac{1\,760}{4\,165}$.

20. $\frac{2}{5}$.

21. Suppose there are A balls, among which a ones are white, in a pot. Now take out balls one by one from the pot without putting back them, until a white ball is taken. The probability to take out a white ball in this way is 1. Then we have

$$\frac{a}{A} + \frac{A-a}{A} \times \frac{a}{A-1} + \frac{(A-a)(A-a-1)}{A(A-1)} \times \frac{a}{(A-2)} + \cdots + \frac{(A-a)(A-a-1)\cdots 1}{A(A-1)\cdots(a+1)} \times \frac{a}{a} = 1.$$

Multiplying both sides by $\frac{A}{a}$, we derive the required identity. This completes the proof.

22. $\frac{10 \times 9 \times 8 \times 7}{10^4} = \frac{63}{125}$.

23. (1) $\frac{C_5^2}{C_{10}^3} = \frac{1}{12}$; (2) $\frac{C_4^2}{C_{10}^3} = \frac{1}{20}$.

24. (1) $\frac{C_{800}^{180} \times C_{2\,200}^{220}}{C_{3\,000}^{400}}$; (2) $1 - \frac{C_{2\,200}^{400} + C_{2\,200}^{399} \times C_{800}^{1}}{C_{3\,000}^{400}}$.

25. $\frac{C_{10}^4 \times C_4^3 \times C_3^2}{C_{17}^9}$.

26. $1 - \frac{C_5^4 \times 2^4}{C_{10}^4} = 1 - \frac{8}{21} = \frac{13}{21}$.

27. The probability that there is no unqualified product is $\frac{C_{95}^{50}}{C_{100}^{50}} = \frac{1\,081}{38\,412}$, and that there is at least one unqualified product is $1 - \frac{1\,081}{38\,412} = \frac{37\,331}{38\,412}$.

28. (1) $1 - \frac{C_4^3}{C_{20}^3} = 1 - \frac{1}{285} = \frac{284}{285}$, (2) $1 - \frac{C_{16}^3}{C_{20}^3} = 1 - \frac{28}{57} = \frac{29}{57}$, (3) $\frac{28}{57}$, and (4) $\frac{1}{285}$.

29. $\frac{1 + 10 + 50}{10\,000} = \frac{61}{10\,000}$.

30. $\frac{3}{20}$.

31. $\frac{1}{C_5^2} = \frac{1}{10}$. If the first ball is put back, the probability is $\left(\frac{2}{5}\right)^2 = \frac{4}{25}$.

32. From $x + \frac{100}{x} > 50$, we find $x \leqslant 2$ or $x \geqslant 48$. Then the required probability is

$$\frac{2 + (100 - 48 + 1)}{100} = \frac{55}{100} = 0.55.$$

33. $\frac{68}{250} = \frac{34}{125}$, $\frac{135 + 68}{250} = \frac{203}{250}$.

34. $\frac{800}{1\,000 + 800} \times 80\% + \frac{1\,000}{1\,000 + 800} \times 70\% = \frac{1\,340}{1\,800} = \frac{67}{90}$.

35. (1) $100 \times 20\% = 20$, i. e. $\frac{C_{20}^4}{C_{100}^4} = \frac{323}{261\,415}$;

(2) $\frac{C_{20}^2 C_{80}^2}{C_{100}^4} = \frac{24\,016}{156\,849}$.

36. $0.84 \times 0.78 = 0.655\,2$.

37. (1) If the plane is shot down by two bullets, then either the second part is shot by two bullets or the first part is shot by at least one bullet. The corresponding probability is then

$$\left(\frac{2}{10}\right)^2 + 1 - \left(\frac{2+7}{10}\right)^2 = 0.23.$$

(2) There is only one possible result that the plane has been shot by three bullets but does not fall down, i.e. the second part is shot by one bullet and the third is shot by two. The corresponding probability is then $1 - C_3^1 \left(\frac{7}{10}\right)^2 \times \frac{2}{10} = 1 - 0.294 = 0.706$.

38. The probability that all the bombs are used up is $0.7^3 \times 0.6^3 = 0.074\,088$. Therefore, the required probability is

$$1 - 0.074\,088 = 0.925\,912.$$

39. The probability that the plane is not destroyed is $(1 - 0.2^2) \times (1 - 0.3)$. Therefore, the probability that it is destroyed is $1 - (1 - 0.2^2) \times (1 - 0.3) = 0.328$.

40. (1) The probability that the plane is hit once is

$$0.4 \times (1 - 0.5) \times (1 - 0.7) + (1 - 0.4) \times 0.5 \times (1 - 0.7) + (1 - 0.4) \times (1 - 0.5) \times 0.7 = 0.36,$$

that it is hit twice is

$$0.4 \times 0.5 \times (1 - 0.7) + (1 - 0.4) \times 0.5 \times 0.7 + 0.4 \times (1 - 0.5) \times 0.7 = 0.41,$$

and that it is hit three times is $0.4 \times 0.5 \times 0.7 = 0.14$. Therefore, the probability that plane is shot down is

$$0.36 \times 0.2 + 0.41 \times 0.6 + 0.14 = 0.458.$$

(2) $\frac{0.36 \times 0.2}{0.458} = \frac{72}{458} \approx 0.157\,2$.

41. $0.05 \times 0.1 + 0.1 \times 0.7 + 0.2 \times 0.2 = 0.115$. The required probability is

$$\frac{0.1 \times 0.7}{0.115} = \frac{70}{115} = \frac{14}{23} \approx 0.608\,7.$$

42. $\frac{0.8 \times (1 - 0.4)}{0.8 \times (1 - 0.4) + (1 - 0.8) \times 0.4} = \frac{48}{56} = \frac{6}{7}$.

43. The probability the plane does not get hit is $(1-0.3)^4$, that it is hit once is $C_4^1\times 0.3\times(1-0.3)^3$. Therefore, the required probability is

$$0.6\times C_4^1\times 0.3\times(1-0.3)^3+(1-C_4^1\times 0.3\times (1-0.3)^3-(1-0.3)^4)=0.595\,26.$$

44. $1-(1-0.2)^8-C_8^1\times 0.2\times(1-0.2)^7\approx 0.496\,7.$

45. There are three cases where the sum of spot numbers is 7: $1+6$, $2+5$, and $3+4$. And each case appears twice. Therefore, the probability is $\frac{2\times 1}{2\times 3}=\frac{1}{3}$.

46. In these 11 letters, "b" and "i" each appear twice, and the other 7 ones each appear once. So the required probability is

$$\frac{1}{11}\times\frac{2}{10}\times\frac{2}{9}\times\frac{1}{8}\times\frac{1}{7}\times\frac{1}{6}\times\frac{1}{5}=0.000\,002\,405.$$

47. When the largest number is 1, the required probability is $\frac{4\times 3\times 2}{4^3}=\frac{3}{8}$; when it is 3, the probability is $\frac{4}{4^3}=\frac{1}{16}$; and when it is 2, the probability is $1-\frac{3}{8}-\frac{1}{16}=\frac{9}{16}$.

48. $\frac{3}{50}\times\frac{2}{49}\times\frac{1}{48}\times 10=\frac{1}{1\,960}$.

49. The probability that A wins is

$$\frac{1}{6}+\frac{5}{6}\times\frac{4}{6}\times\frac{3}{6}+\frac{5}{6}\times\frac{4}{6}\times\frac{3}{6}\times\frac{2}{6}\times\frac{5}{6}=\frac{169}{324},$$

and that B wins is

$$\frac{5}{6}\times\frac{2}{6}+\frac{5}{6}\times\frac{4}{6}\times\frac{3}{6}\times\frac{4}{6}+\frac{5}{6}\times\frac{4}{6}\times\frac{3}{6}\times\frac{2}{6}\times\frac{1}{6}=\frac{155}{324}.$$

50. There are 5 sets of 7 digits whose sum is 59, as shown in the following: (1) six 9s, one 5; (2) five 9s, one 8, one 6; (3) five 9s, two 7s; (4) four 9s, two 8s, one 7; and (5) three 9s, four 8s. Their total permutations are

$$7+7\times 6+C_7^2+7\times C_6^2+C_7^4=210.$$

A number is divisible by 11, if and only if the difference between the sum of digits at even places and that at odd places is divisible by 11. As the total sum is 59, then the difference must be 11. So the sum of 4 digits at odd places is 35 $\left(=\frac{59+11}{2}\right)$ and that of 3 digits at even places is 24 $\left(=\frac{59-11}{2}\right)$.

In the meantime, there are only one subset of 4 digits at odd places whose sum is 35, which are three 9s and one 8, having 4 permutations.

And there are three subsets of 3 digits at even places whose each sum is 24, which are (1) three 8s, having one permutation; (2) two 9s and one 6, having three permutations; and (3) one 9, one 8 and one 7, having $3!=6$ permutations. The total permutations are then

$$1+3+3!=10.$$

Therefore, the number of the numbers that are divisible by 11 is $4\times 10=40$.

The required probability is then

$$\frac{40}{210}=\frac{4}{21}.$$

51. There are two sets of three digits whose sum is 10 in the first case: {2, 3, 5} and {1, 4, 5}. The corresponding probability is

$$\frac{2}{C_{10}^3}=\frac{1}{60}.$$

There are 5 sets of 3 digits whose sum is 10 in the second case: {0, 5, 5}, {1, 4, 5}, {2, 3, 5}, {2, 4, 4} and {3, 3, 4}, and they have 33 permutations. Therefore, the corresponding probability is

$$\frac{33}{10^3}=\frac{33}{1\,000}.$$

52. There are 10 digits (i.e. 0, 1, 2, 3, 4, 5, 6, 7, 8, 9 and 10) that may appear at the ones place of a number. If the ones place of a product is 1, 3, 7 or 9, then that of each multiplier of it is also 1, 3, 7 or 9. The required probability is then $\left(\frac{4}{10}\right)^n = \frac{2^n}{5^n}$.

If the ones place of a product is 2, 4, 6 or 8, then that of each multiplier of it is neither 0 nor 5, and at least one of them is an even number. Therefore, the probability is $\frac{8^n - 4^n}{10^n} = \frac{4^n - 2^n}{5^n}$.

If the ones place of a product is 5, then that of each multiplier of it is an odd number, and at least one of them is 5. Therefore, the probability is $\frac{5^n - 4^n}{10^n}$.

If the ones place of a product is 0, then the required probability is

$$1 - \frac{2^n}{5^n} - \frac{4^n - 2^n}{5^n} - \frac{5^n - 4^n}{10^n} = \frac{10^n - 8^n - 5^n + 4}{10^n}.$$

53. There are totally 3^{10} ways of allocation. Among them, there is only one way that A gets 10 things, there are $C_{10}^9 \times 2$ ways he gets 9 ones, $C_{10}^8 \times 2^2$ ways he gets 8 ones, $C_{10}^7 \times 2^3$ ways he gets 7 ones, and $C_{10}^6 \times 2^4$ ways he gets 6 ones. Therefore, the required probability is

$$\frac{1 + C_{10}^9 \times 2 + C_{10}^8 \times 2^2 + C_{10}^7 \times 2^3 + C_{10}^6 \times 2^4}{3^{10}} = \frac{1\,507}{19\,683}.$$

54. $P(AB) = P(A)P(B \mid A) = \frac{1}{6} \times \frac{1}{2} = \frac{1}{12}$, $P(B) = \frac{P(AB)}{P(A \mid B)} = \frac{1}{12} \div \frac{1}{3} = \frac{1}{4}$; therefore,

$$P(A \cup B) = P(A) + P(B) - P(AB) = \frac{1}{6} + \frac{1}{4} - \frac{1}{12} = \frac{1}{3}.$$

55. $P(\bar{B} \mid \bar{A} \cup B) = \frac{P(\bar{B}(\bar{A} \cup B))}{P(\bar{A} \cup B)} = \frac{P(\bar{B}\bar{A})}{P(\bar{A} \cup B)}$, $P(\bar{B}\bar{A}) = P(\bar{B}) - P(\bar{B}A) = 0.6 - 0.5 = 0.1$.

Since $\overline{\bar{A} \cup B} = A\bar{B}$, then

$$P(\bar{A} \cup B) = 1 - P(A\bar{B}) = 1 - 0.5 = 0.5.$$

Therefore, $P(\bar{B} \mid \bar{A} \cup B) = \frac{0.1}{0.5} = \frac{1}{5}$.

56. A and B are connected if and only if switches 1 and 3 are closed or switches 2 and 3 are closed. Therefore, the required probability is

$$p(p + p - p \cdot p) = p^2(2 - p), \text{ or } p(1 - (1 - p)^2) = p^2(2 - p).$$

57. The required probability is

$$p(p + p^2 - p \cdot p^2) = p^2(1 + p - p^2),$$

or

$$p(1 - (1 - p) \cdot (1 - p^2)) = p^2(1 + p - p^2).$$

58. Let A_i ($i = 1, 2, 3, 4, 5$) denote the event that switch i is closed. Then the required probability is

$$\begin{aligned} & P(A_1A_2 \cup A_4A_5 \cup A_1A_3A_5 \cup A_2A_3A_4) \\ &= P(A_1A_2) + P(A_4A_5) + P(A_1A_3A_5) + P(A_2A_3A_4) - \\ & P(A_1A_2A_4A_5) - P(A_1A_2A_3A_5) - P(A_1A_2A_3A_4) - \\ & P(A_1A_3A_4A_5) - P(A_2A_3A_4A_5) - P(A_1A_2A_3A_4A_5) + \\ & 4P(A_1A_2A_3A_4A_5) - P(A_1A_2A_3A_4A_5) \\ &= 2p^2 + 2p^3 - 5p^4 + 2p^5. \end{aligned}$$

59. $\frac{1}{10}$ (See Chapter 50).

60. $\frac{a}{a+b} \times \frac{c+1}{c+d+1} + \frac{b}{a+b} \times \frac{c}{c+d+1} = \frac{ac+a+bc}{(a+b)(c+d+1)}$.

61. The probability to dial the phone successfully by just four tries is $\frac{9}{10} \times \frac{8}{9} \times \frac{7}{8} \times \frac{1}{7} = \frac{1}{10}$, and that to succeed with at most four tries is $\frac{4}{10} = \frac{2}{5}$.

62. $1 - \left(1 - \frac{1}{3}\right)\left(1 - \frac{1}{4}\right)\left(1 - \frac{1}{5}\right) = 1 - \frac{2}{5} = \frac{3}{5}$.

63. $$\frac{\frac{10}{15} \times \frac{1}{2^3}}{\frac{10}{15} \times \frac{1}{2^3} + \frac{5}{15} \times 1} = \frac{10}{10 + 5 \times 2^3} = \frac{1}{5}.$$

64. (1) $\frac{8}{10}\times\frac{7}{9}$, (2) $\frac{8}{10}\times\frac{2}{9}+\frac{2}{10}\times\frac{8}{9}=\frac{16}{45}$, and (3) $\frac{2}{10}=\frac{1}{5}$.

65. The probability that both the child and mother get the flu is $0.4\times 0.5=0.2$. Then the required probability is

$$0.2\times(1-0.4)=0.12.$$

66. The required probability is

$$37.5\%(1-20\%)+20\%(1-37.5\%)+$$
$$(1-37.5\%-20\%-35\%)+35\%\times 35\%=62.25\%.$$

67.
$$\frac{100}{200}\times 5\%+\frac{100}{200}\times 0.25\%=\frac{5.25}{200},$$
$$\frac{100}{200}\times 5\%\div\frac{5.25}{200}=\frac{500}{525}=\frac{20}{21}.$$

68. The probability for inputting *AAAA* to get the output *ABCA* is $\alpha^2\left(\frac{1-\alpha}{2}\right)^2$, that for inputting *BBBB* to get *ABCA* is $\alpha\left(\frac{1-\alpha}{2}\right)^3$, and that for inputting *CCCC* to get *ABCA* is $\alpha\left(\frac{1-\alpha}{2}\right)^3$. Therefore, the probability that the output *ABCA* is produced by the input *AAAA* is

$$\frac{p_1\times\alpha^2\left(\frac{1-\alpha}{2}\right)^2}{p_1\times\alpha^2\left(\frac{1-\alpha}{2}\right)^2+p_2\times\alpha\left(\frac{1-\alpha}{2}\right)^3+p_3\times\alpha\left(\frac{1-\alpha}{2}\right)^3}$$
$$=\frac{2\alpha p_1}{(3\alpha-1)p_1+1-\alpha}.$$

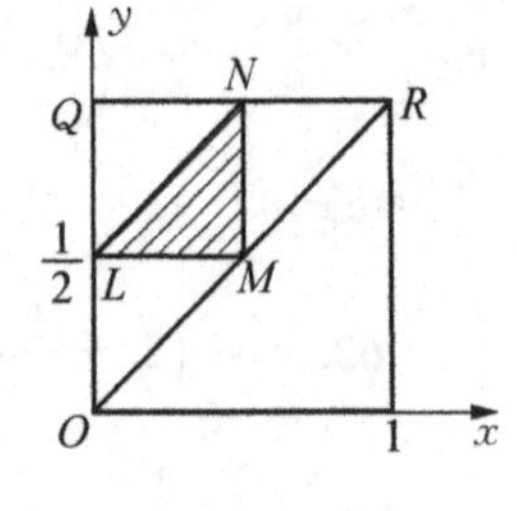

69. Let the three points be A, B and C, cut the circle at point A, and straighten it into a line segment. Then the original problem turns into this: taking two points B and C one after one on a unit segment AA', find the probability that $AB<\frac{1}{2}$, $CA'<\frac{1}{2}$ and $BC<\frac{1}{2}$. Let $AB=x$, $AC=y$. Then points (x, y) $(0\leqslant x<y\leqslant 1)$ constitute $\triangle ORQ$ in the

coordinate plane (see the figure), while points satisfying $x < \frac{1}{2}$, $1 - y < \frac{1}{2}$ and $y - x < \frac{1}{2}$ constitute $\triangle LMN$ (the shaded region in the figure). It is obvious that the ratio of the area of $\triangle LMN$ to that of $\triangle ORQ$ is $\frac{1}{4}$. Therefore, the required probability is $\frac{1}{4}$.

70. At this time, the three points constitute an acute triangle. Therefore, it goes back to the last problem. So the probability is also $\frac{1}{4}$.

71. Suppose the distances of two points from one end of the segment are x, y $(x < y)$, respectively. Then points (x, y) constitute the large triangle in the figure right, while that satisfying $y - x > b$ constitute the small triangle (the shaded region in the figure). The required probability is the ratio of the areas of the two triangles, and that is

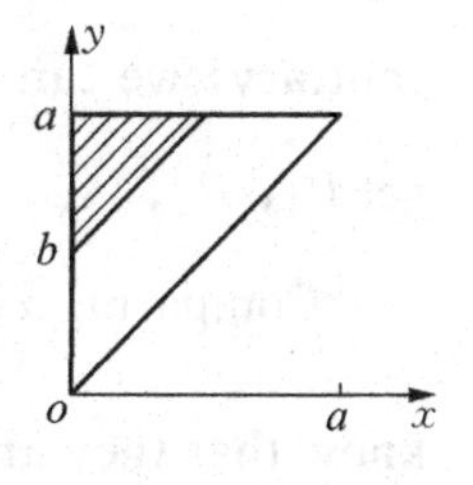

$$\frac{1}{2}(a-b)^2 \div \frac{1}{2}a^2 = \left(\frac{a-b}{a}\right)^2.$$

72. Referring to the solution of Exercise 69, the three parts constitute a triangle if and only if $x < \frac{1}{2}$, $1 - y < \frac{1}{2}$ and $y - x < \frac{1}{2}$. Therefore, the required probability is $\frac{1}{4}$.

73. We may assume that the length of the rod is 1. Denote the n points satisfying the condition as P_1, P_2, ..., P_n, with their distances to one end of the rod being $x_1 \leqslant x_2 \leqslant \cdots \leqslant x_n \leqslant 1$, respectively. In addition, suppose points A_0, A_1, A_2, ..., A_n divide the rod into n equal parts of length $\frac{1}{n}$, with A_0 at the rod's same end mentioned above and A_n at the other end. Then, each equal part must contain exactly one of P_1, P_2, ..., P_n; otherwise, the length of some segment will be greater than $\frac{1}{n}$. Furthermore, we have $A_0P_1 \geqslant A_1P_2$;

otherwise

$$P_1P_2 = P_1A_1 + A_1P_2 > P_1A_1 + A_0P_1 = A_0A_1 = \frac{1}{n}.$$

By the same reason, we have $A_1P_2 \geqslant A_2P_3 \geqslant \cdots \geqslant A_{n-1}P_n$.

Now, move the second, third, etc., and the nth equal parts together to the first part, letting A_1, A_2, ..., A_n coincide with A_0. Then P_1, P_2, ..., P_n become the points in A_0A_1 (the first equal part). We denote their distances from the end A_0 as $y_n (= x_1)$, y_{n-1}, ..., y_1, respectively, then we have $y_1 \leqslant y_2 \leqslant \cdots \leqslant y_n \leqslant \frac{1}{n}$. In the contrary, we can also from a group of points $y_1 \leqslant y_2 \leqslant \cdots \leqslant y_n \leqslant \frac{1}{n}$ get P_1, P_2, ..., P_n satisfying the required condition.

Comparing $x_1 \leqslant x_2 \leqslant \cdots \leqslant x_n \leqslant 1$ to $y_1 \leqslant y_2 \leqslant \cdots \leqslant y_n \leqslant \frac{1}{n}$, we know that they are one-to-one correspondence by multiplying n $\left(\frac{1}{n}\right)$.

Therefore, the required probability is $\frac{1}{n^n}$.

74. Let the segment $AB = a$, and the selected points C, D satisfy $AC = x$ and $DB = y$, respectively, as shown in Figure (1) below. Then $0 < x$, $y < a$ and $x + y < 1$, which constitute $\triangle OPQ$ in Figure (2).

When $b \geqslant \frac{a}{2}$, points (x, y) satisfying $0 < x < b$, $0 < y < b$ and $a - (x + y) < b$ constitute the shaded region in Figure (2). The ratio of the area of the region to that of $\triangle OPQ$ is

$$1 - 3\left(\frac{a-b}{a}\right)^2,$$

which is also the required probability.

When $b < \frac{a}{2}$, points (x, y) satisfying $0 < x < b$, $0 < y < b$ and $a - (x + y) < b$ constitute the shaded right triangle in Figure (3), whose side is $b - (a - b - b) = 3b - a$. The required probability is then

$\left(\frac{3b-a}{a}\right)^2.$

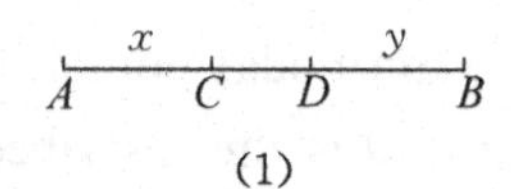

(1)

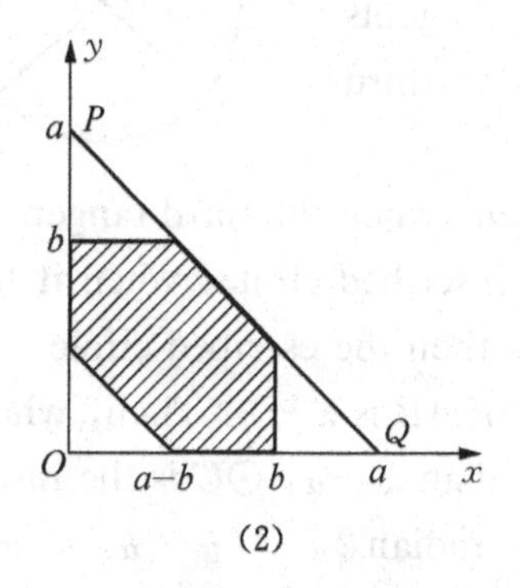

(2)

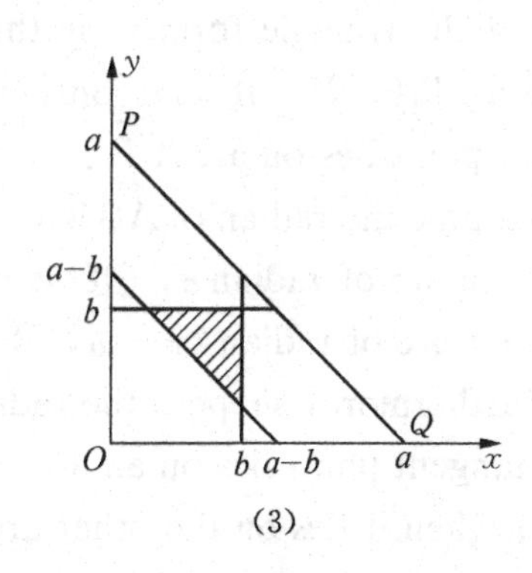

(3)

75. Let the two remaining segments be x and y, respectively. Then points (x, y) satisfying $0 \leqslant x \leqslant a$ and $0 \leqslant y \leqslant a$ form a square in the coordinate plane as shown in the figure.

(1) $x + y > a$ is the triangle above the diagonal line, whose area is $\frac{1}{2}$ of that of the square. So the required probability is $\frac{1}{2}$.

(2) $x + y < a$ is the triangle below the diagonal line, whose area is also $\frac{1}{2}$ of that of the square. So the required probability is $\frac{1}{2}$ too.

(3) $x + y = a$ is the diagonal line, whose area is zero. Therefore, the corresponding probability is 0.

76. Suppose the length of the longest segment among the three is a, and the lengths of the other two are x and y, respectively. Then we have $x + y > a$, whose probability is $\frac{1}{2}$ as is proved in the exercise above. Therefore, the required probability is $\frac{1}{2}$.

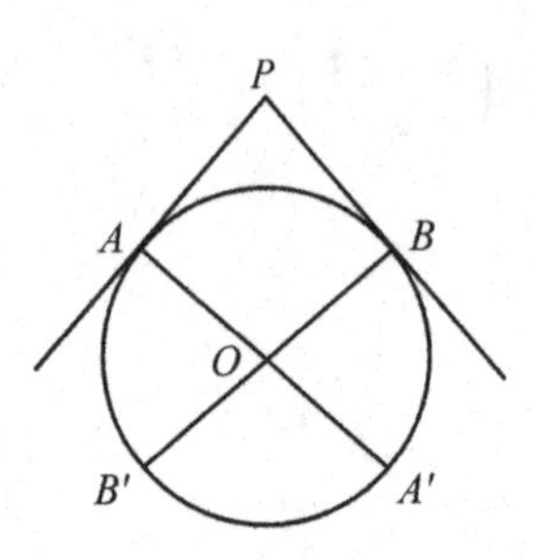

77. Let PA, PB be two tangents to $\odot O$ at points A, B, respectively, as seen in the figure. Let A', B' be the antipodal points of A, B, respectively. Then $\odot O$ is the inscribed circle of the triangle formed by three tangents including PA, PB if and only if the third tangent point lies on arc $\overset{\frown}{A'B'}$.

Suppose the radian of $\overset{\frown}{AB}$ is α. Then, when the third tangent point lies on an arc of radian α, $\odot O$ is the inscribed circle; when it lies on the other arc of radian $2\pi - \alpha$, $\odot O$ is then the escribed circle.

Furthermore, suppose the radian of $\overset{\frown}{AB}$ is $\pi - \alpha$. Then, when the third tangent point lies on an arc of radian $\pi - \alpha$, $\odot O$ is the inscribed circle; when it lies on the other arc of radian $2\pi - (\pi - \alpha) = \pi + \alpha$, $\odot O$ is then the escribed circle.

Combining two cases above, we have the radian of the arc on which the third tangent point lies to make $\odot O$ the inscribed circle is $\alpha + (\pi - \alpha) = \pi$, while that to make it the escribed circle is $(2\pi - \alpha) + (\pi + \alpha) = 3\pi$.

Since the results obtained above holds for any pair of arcs, the corresponding ratio is then $1 : 3$.

78. As shown in Figures (1) and (2), let two segments be $CD = a$ and $EF = b$ in respect, and let $AC = x$ and $AE = y$. Then $0 \leqslant x \leqslant b$ and $0 \leqslant y \leqslant a$.

(1) When the common part is not greater than c, we have

$$x + a - y < c \text{ or } y + b - x < c.$$

They constitute two shaded triangles in Figure (3), whose area sum is c^2. Therefore, the required probability is $\frac{c^2}{ab}$.

(2) When the shorter segment lies completely in the longer one, we have

$$x < y \text{ and } a - y > b - x.$$

They constitute the shaded parallelogram in Figure (4), whose area is

$b(a-b)$. Therefore, the required probability is $\frac{a-b}{a}$.

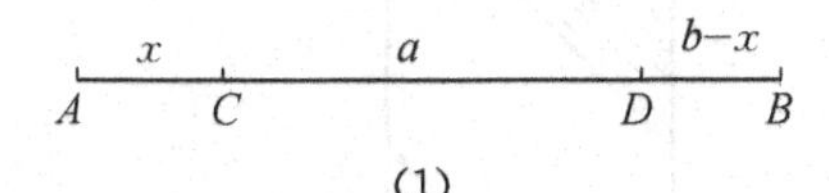

(1)

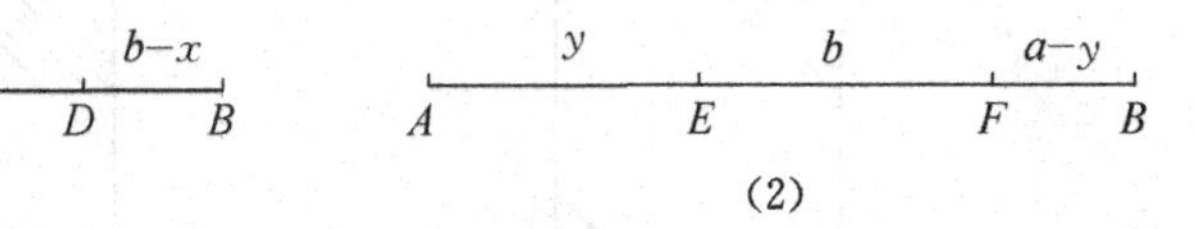

(2)

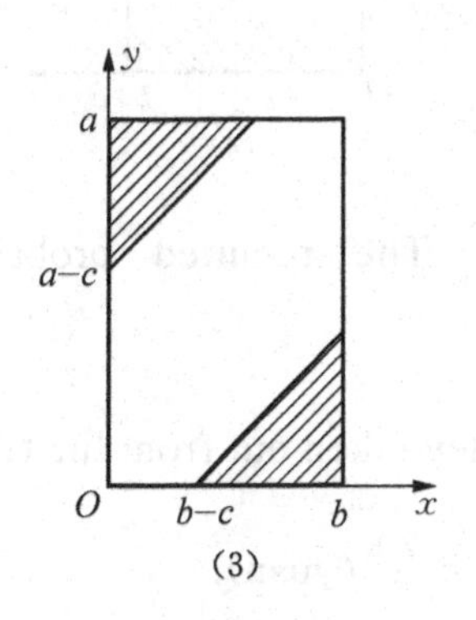

(3)

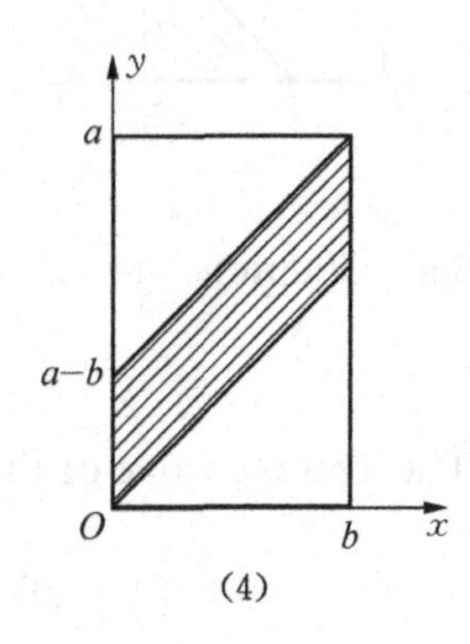

(4)

79. As shown in Figures (1) and (2), let two segments be $CD=a$ and $EF=b$ in respect, and let $AC=x$ and $AE=y$. Then $0<x<b+c$ and $0<y<a+c$.

(1) When two segments have no common part, we have

$$y>a+x \text{ or } x>y+b.$$

They constitute two shaded triangles in Figure (3), whose area sum is c^2. Therefore, the required probability is $\frac{c^2}{(a+c)(b+c)}$.

(2) When the common part is not greater than d, we have

$$y+d>a+x \text{ or } x+d>y+b.$$

They constitute the shaded regions in Figure (4), whose area sum is $(c+d)^2$. Therefore, the required probability is $\frac{(c+d)^2}{(c+a)(c+b)}$.

x a
A C D B

(1)

y b
A E F B

(2)

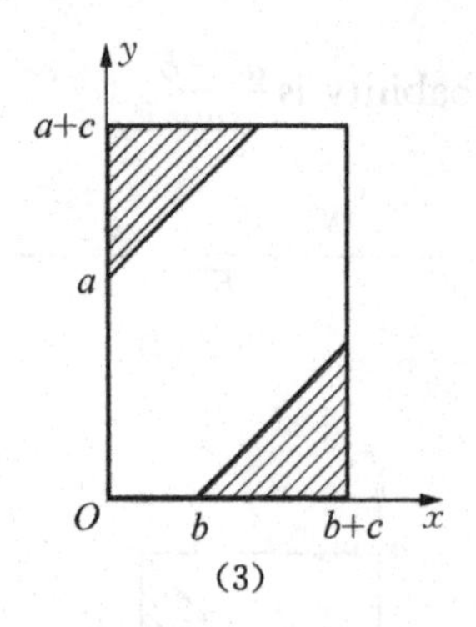

(3)

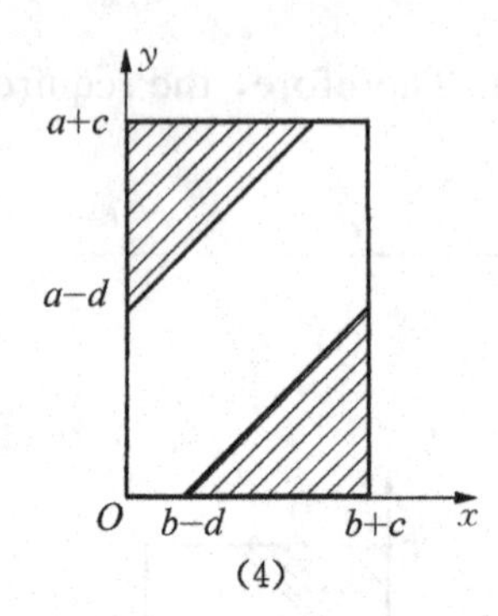

(4)

80. See Chapters 44 and 65. The required probability is $\frac{a+b+c}{\pi d}$.

81. The average value of the money taken out from the first bag is

$$(5 \times 10 + 20) \times \frac{2}{6} = \frac{70}{3} \text{ (yuan)}.$$

The average value of the money taken out from the second bag is

$$\left(6 \times 10 + \frac{70}{3}\right) \times \frac{2}{8} = \frac{125}{6} \text{ (yuan)}.$$

Therefore, the mathematical expectation of the money in the first bag is

$$5 \times 10 + 20 - \frac{70}{3} + \frac{125}{6} = 67\frac{1}{2} \text{(yuan)},$$

and that in the second bag is

$$5 \times 10 + 20 + 6 \times 10 - 67\frac{1}{2} = 62\frac{1}{2} \text{ (yuan)}.$$

Another solution: The probability that the 20-yuan note remains in the first bag is

$$\frac{4}{6} + \frac{2}{6} \times \frac{2}{8} = \frac{3}{4}.$$

Then the expected value of the money in the first bag is

$$5 \times 10 + 20 \times \frac{3}{4} + 10 \times \frac{1}{4} = 67\frac{1}{2} \text{ (yuan)},$$

and that in the second bag is

$$5 \times 10 + 20 \times \frac{1}{4} + 10 \times \frac{3}{4} = 62\frac{1}{2} \text{ (yuan)}.$$

82. The probability for A to win is

$$\frac{1}{6} + \left(\frac{5}{6}\right)^2 \times \frac{1}{6} + \left(\frac{5}{6}\right)^4 \times \frac{1}{6} + \cdots,$$

and that for B to win is

$$\frac{5}{6} \times \frac{1}{6} + \left(\frac{5}{6}\right)^3 \times \frac{1}{6} + \left(\frac{5}{6}\right)^5 \times \frac{1}{6} + \cdots,$$

which is $\frac{5}{6}$ of the former. Therefore, the expected value for A is

$$110 \times \frac{5}{6+5} = 60 \text{ (yuan)},$$

and that for B is 50 yuan.

83. There are $C_5^2 = 10$ ways to take out 2 notes from the bag, among which there are $C_3^2 = 3$ ways to take two 50-yuan ones, $C_2^1 \times C_3^1 = 6$ ways to take one 100-yuan and one 50-yuan notes, and one way to take two 100-yuan notes. Therefore, the expected value is

$$\frac{1}{10} \times 100 \times 2 + \frac{3}{10} \times 50 \times 2 + \frac{6}{10} \times (50 + 100) = 140 \text{ (yuan)}.$$

A more simple solution: As each note has probability of $\frac{2}{5}$ to be taken out, the expected value is then

$$\frac{2}{5} \times (2 \times 100 + 3 \times 50) = 140 \text{ (yuan)}.$$

84. $\frac{1}{2} \times \frac{2}{5} \times (100 + 4 \times 10) = 28$ (yuan).

85. See Chapter 49. The probability for A to win is

$$\sum_{i=3}^{4} C_4^i \left(\frac{1}{2}\right)^4 = \frac{5}{16}.$$

Then A is expected to get $160 \times \frac{5}{16} = 50$ (yuan), and B is $160 - 50 = 110$ (yuan).

86. The probability to get 7 spots by rolling two dice once is $\frac{6}{36} = \frac{1}{6}$, and that to get 4 spots is $\frac{3}{36} = \frac{1}{12}$. The former is twice of the latter. Let x be the probability that B wins. Then that A wins is $2x$.

$$x + 2x = 1.$$

Therefore $x = \frac{1}{3}$. The mathematical expectation of B is

$$5 \times \frac{1}{3} - 2 \times \frac{2}{3} = \frac{1}{3} \text{ (yuan)}.$$

87. Suppose there are totally n notes. Referring to Chapter 50, the probability to take out the M-yuan note at the kth time is $\frac{1}{n}$. Therefore, the mathematical expectation of the money taken out is

$$\frac{1}{n}M + \frac{1}{n}(M+n) + \frac{1}{n}(M+2n) + \cdots + \frac{1}{n}(M+(n-1)m)$$
$$= M + \frac{1}{n} \cdot \frac{1}{2}n(n-1)m = M + \frac{n-1}{2} \cdot m.$$

88. The mathematical expectation is

$$E = p + 2pq + 3pq^2 + \cdots,$$

where $q = 1 - p$. We have

$$qE = pq + 2pq^2 + 3pq^3 + \cdots.$$

The former expression minus the latter one, we get

$$pE = p + pq + pq^2 + pq^3 + \cdots = p \times \frac{1}{1-q} = 1.$$

Therefore, $E = \dfrac{1}{p}$.

89.

$$\sum_{k=0}^{n} k^2 C_n^k p^k q^{n-k} = \sum_{k=0}^{n} k(k-1) C_n^k p^k q^{n-k} + np$$

$$= \sum_{k=2}^{n} n(n-1) C_{n-2}^{k-2} p^k q^{n-k} + np$$

$$= n(n-1)p^2 \sum_{k=0}^{n-2} C_{n-2}^k p^k q^{n-k} + np$$

$$= n(n-1)p^2 + np = n^2p^2 + npq,$$

$$\sum_{k=0}^{n} (k-np)^2 C_n^k p^k q^{n-k} = \sum_{k=0}^{n} k^2 C_n^k p^k q^{n-k} - 2np \sum_{k=0}^{n} k C_n^k p^k q^{n-k} + n^2p^2$$

$$= n^2p^2 + npq - 2n^2p^2 + n^2p^2 = npq.$$

90. The probability for a given day to be exactly k students' birthday is

$$C_n^k \times \left(\frac{1}{365}\right)^k \times \left(\frac{364}{365}\right)^{n-k}.$$

Referring to Chapter 51 $\left(\text{let } n = 365 \text{ and } p = C_n^k \times \left(\frac{1}{365}\right)^k \times \left(\frac{364}{365}\right)^{n-k}\right)$, the mathematical expectation of x is

$$365 \times C_n^k \times \left(\frac{1}{365}\right)^k \times \left(\frac{364}{365}\right)^{n-k} = C_n^k \times \frac{364^{n-k}}{365^{n-1}}.$$

91. $365\left(1 - \left(\frac{364}{365}\right)^n - n \times \left(\frac{364}{365}\right)^{n-1} \times \frac{1}{365}\right)$.

92. According to Exercise 89, we have

$$npq = \sum_{k=0}^{n} (k-np)^2 C_n^k p^k q^{n-k} = n^2\varepsilon^2 \sum_{k=0}^{n} \left(\frac{k-np}{n\varepsilon}\right)^2 C_n^k p^k q^{n-k}$$

$$> n^2\varepsilon^2 \sum_{|k-np|>n\varepsilon} C_n^k p^k q^{n-k} = n^2\varepsilon^2 P\left(\left|\frac{m}{n} - p\right| > \varepsilon\right).$$

Therefore

$$P\left(\left|\frac{m}{n} - p\right| > \varepsilon\right) < \frac{npq}{n^2\varepsilon^2} = \frac{p(1-p)}{n\varepsilon^2} \leqslant \frac{1}{4n\varepsilon^2}.$$

This exercise is a special case of the famous Chebyshev's inequality.

93. $$\sum_{k=0}^{n} k^3 C_n^k p^k q^{n-k}$$

$$= \sum_{k=0}^{n} k(k-1)(k-2)C_n^k p^k q^{n-k} + 3\sum_{k=0}^{n} k^2 C_n^k p^k q^{n-k} - 2\sum_{k=0}^{n} k C_n^k p^k q^{n-k}$$

$$= \sum_{k=0}^{n} k(k-1)(k-2)C_n^k p^k q^{n-k} + 3(n^2p^2 + npq) - 2np$$

$$= \sum_{k=0}^{n-3} k(k-1)(k-2)C_{n-3}^k p^k q^{n-3-k} \cdot n \cdot (n-1)(n-2)p^3 + 3(n^2p^2 + npq) - 2np$$

$$= n(n-1)(n-2)p^3 + 3n^2p^2 - 3np^2 + np.$$

94. $$E[(X-EX)^2] = E(X^2 - 2X \cdot EX + (EX)^2) = E(X^2) - (EX)^2 \leqslant E(X^2).$$

The equality holds if and only if $EX = 0$.

95. The required probability is

$$\frac{C_n^k C_{N-n}^{r-k}}{C_N^r} = \frac{C_n^k \cdot r!}{(r-k!)} \times \frac{(N-n)!(N-r)!}{N!(N-n-r+k)}.$$

When N becomes $N+1$, the right side of the expression will be multiplied by a factor $\frac{(N+1-n)(N+1-r)}{(N+1)(N+1-n-r+k)}$. As

$$\frac{(N+1-n)(N+1-r)}{(N+1)(N+1-n-r+k)} \leqslant 1 \Leftrightarrow nr \leqslant k(N+1) \Leftrightarrow \frac{nr}{k} \leqslant N+1,$$

we have that: The probability reaches the maximum when $N = \left[\frac{nr}{k}\right]$.

96. We first answer question (2). Let the probabilities for A, B, C to win the competition be p_1, p_2, p_3, respectively. It is obvious that

$$p_1 + p_2 + p_3 = 1. \tag{1}$$

(There must be a champion, as the probability for the competition to continue indefinitely is $\frac{1}{2} \times \frac{1}{2} \times \frac{1}{2} \times \cdots = 0$.)

There are two possible ways for A to win the competition: He wins C in the second round with probability $\frac{1}{2}$; or he loses to C in this round with probability $\frac{1}{2}$, and then he is in the same state as B was before the second round. Therefore

$$p_1 = \frac{1}{2} + \frac{1}{2}p_2. \tag{2}$$

In order to win the competition, C must win A in the second round with probability $\frac{1}{2}$, and then he will be in the same state as A before the second round. Therefore

$$p_3 = \frac{1}{2}p_1. \tag{3}$$

From expressions (1)-(3), we get $p_1 = \frac{4}{7}$, $p_2 = \frac{1}{7}$, $p_3 = \frac{2}{7}$.

As for question (1), the probabilities for A and B to win the champion is the same as

$$\frac{1}{2}p_1 + \frac{1}{2}p_2 = \frac{5}{14},$$

and that for C is $1 - 2 \times \frac{5}{14} = \frac{2}{7}$.

97. There are $C_6^3 = 20$ ways selecting three points from six ones to form $\triangle ABC$. Among them, there are $C_6^1 = 6$ ways in which the three points are adjacent one by one, and at this time the sides of $\triangle ABC$ will not cross that of the triangle formed by the other three points.

So the required probability is $\frac{6}{20} = \frac{3}{10}$.

98. We may assume that the integers on each side is 0, 1 or 2. (If it is not, replace it with the remainder of dividing it by 3.) Furthermore, suppose there are a 0s, b 1s, and c 2s on the sides of each dice, where $0 \leqslant a, b, c < n$ and $a + b + c = n$.

There are n^3 possible results in rolling three dice randomly. Among them, the sets of integers facing up with sums divisible by 3 are

$$\{0, 0, 0\}, \{1, 1, 1\}, \{2, 2, 2\}, \{0, 1, 2\},$$

which account for

$$a^3 + b^3 + c^3 + 6abc$$

possible results. Therefore, the required probability is $\dfrac{a^3 + b^3 + c^3 + 6abc}{n^3}$.

$$\frac{a^3 + b^3 + c^3 + 6abc}{n^3} \geqslant \frac{1}{4}$$
$$\Leftrightarrow 4(a^3 + b^3 + c^3 + 6abc) \geqslant (a + b + c)^3$$
$$\Leftrightarrow a^3 + b^3 + c^3 + 6abc \geqslant a^2b + a^2c + b^2a + b^2c + c^2a + c^2b.$$

We may assume that $a \geqslant b \geqslant c$. Then we have

$$\begin{aligned}
& a^3 + b^3 + 2abc - (a^2b + a^2c + b^2a + b^2c) \\
&= a^2(a - b) - b^2(a - b) - ac(a - b) + bc(a - b) \\
&= (a - b)(a^2 - b^2 - ac + bc) \\
&= (a - b)^2(a + b - c) \geqslant 0;
\end{aligned}$$

$$c^3 + abc - c^2a - c^2b = c^2(c - a) + bc(a - c) = c(a - c)(b - c) \geqslant 0.$$

Adding two expressions above, then we get the required result and finish the proof.

99. Let the probability of getting n points be P_n. The event that we cannot get n points occurs only when we get $n - 1$ points first and then toss out the tail side facing up. Therefore

$$1 - P_n = \frac{1}{2}P_{n-1},$$

or

$$\frac{2}{3} - P_n = \frac{1}{2}\left(P_{n-1} - \frac{2}{3}\right).$$

Then we have

$$\begin{aligned}
\frac{2}{3} - P_n &= \left(-\frac{1}{2}\right)^{n-1}\left(\frac{2}{3} - P_1\right) \\
&= \left(-\frac{1}{2}\right)^{n-1}\left(\frac{2}{3} - \frac{1}{2}\right) \\
&= -\frac{1}{3}\left(-\frac{1}{2}\right)^n.
\end{aligned}$$

So

$$P_n = \frac{2}{3} + \frac{1}{3}\left(-\frac{1}{2}\right)^n = \frac{2 + \left(-\frac{1}{2}\right)^n}{3}.$$

100. For each case where the kth card is the second ace, we inverse the order of the cards in the pack and then have a corresponding case where the $(N - k + 1)$th card is the second ace. Therefore, we are able to pair all the cases (except for the case $\frac{N+1}{2}$, which does not need to be paired), with the average number of cards taken for each pair is $\frac{N+1}{2}$. Consequently, the required mathematical expectation is $\frac{N+1}{2}$. The proof is finished. (This is essentially the problem in Chapter 50.)

101. Referring to Chapter 44, the total number of balls is $t = \frac{x_4 + 1}{2} = 120$, and the number of red balls is $r = \frac{y_4 + 1}{2} = 85$.

102. Suppose the ith jar contains x_i male and y_i female turtles $(1 \leqslant i \leqslant 5)$. Then, when we randomly take out a turtle from these jars, the probability for it being male is

$$P = \frac{1}{5}\sum_{i=1}^{5}\frac{x_i}{x_i + y_i}.$$

We need a proposition (it will be proved later): Suppose a_1, b_1, a_2, b_2 are positive integers, satisfying $a_1 \leqslant b_1$, $a_2 \leqslant b_2$. Then

$$\frac{a_1}{b_1} + \frac{a_2}{b_2} \leqslant 1 + \frac{a_1 + a_2 - 1}{b_1 + b_2 - 1}.$$

By using the proposition repeatedly, we get

$$P \leqslant \frac{1}{5}\left(1 + 1 + 1 + 1 + \frac{x_1 + x_2 + x_3 + x_4 + x_5 - 4}{x_1 + x_2 + x_3 + x_4 + x_5 + y_1 + y_2 + y_3 + y_4 + y_5 - 4}\right)$$

$$= \frac{1}{2}\left(4 + \frac{60 - 4}{160 - 4}\right) = 0.871\,79\ldots.$$

Therefore, we should put one male turtle into each of the first four jars, respectively, and put all the other turtles into the last jar, to get the greatest probability required as 0.871 79....

(Proof of the proposition: We have

$$1+\frac{a_1+a_2-1}{b_1+b_2-1}-\left(\frac{a_1}{b_1}+\frac{a_2}{b_2}\right)$$
$$=\frac{(b_1^2-b_1)(b_2-a_2)+(b_2^2-b_2)(b_1-a_1)}{b_1b_2(b_1+b_2-1)}\geqslant 0.$$

This completes the proof.)

103. Think a horse stays in place $x=1$ of the number axis. He goes toward the origin one step with probability p, and goes one step in the opposite direction with probability q. According to Chapter 35, the probability that he reaches the origin finally is $\frac{p}{q}$ (if $p<\frac{1}{2}$) or 1 (if $p\geqslant\frac{1}{2}$).

On the other hand, assume that he reaches the origin for the first time after $2k+1$ steps. Then his last step is from $x=1$ to the origin; before that, he has k steps in the positive direction of the number axis and k steps toward the origin — during this process, the steps in the positive direction is always not less than that toward the origin. Then, referring to the money change problem in Chapter 26, there are $\frac{C_{2k}^k}{k+1}$ walking lines with probability $\frac{C_{2k}^k}{k+1}p^{k+1}q^k$ for the horse. Therefore,

$$\sum_{k=0}^{\infty}\frac{C_{2k}^k}{k+1}p^{k+1}q^k=\frac{p}{q}\text{ or 1 (according to whether }p<\frac{1}{2}\text{ or }p\geqslant\frac{1}{2}\text{).}$$

Particularly, it equals $\frac{1}{2}$ for $p=\frac{1}{3}$.

104. Fishing n times, the probability for it not to be caught is $\left(1-\frac{1}{k}\right)^n$, and the gaining is nm yuan. Therefore, the mathematical expectation is $nm\left(1-\frac{1}{k}\right)^n$. We have

$$\frac{(n+1)\left(1-\frac{1}{k}\right)^{n+1}}{n\left(1-\frac{1}{k}\right)^{n}}=\left(1-\frac{1}{k}\right)\left(1+\frac{1}{n}\right)=1+\frac{k-1-n}{kn}.$$

When $n \leqslant k-1$, $(n+1)\left(1-\frac{1}{k}\right)^{n+1} \geqslant n\left(1-\frac{1}{k}\right)^{n}$. Therefore, the gaining reaches the maximum for $k-1 \leqslant n \leqslant k$.

105. Let the probability for us to get n points is P_n. The event that we cannot get n points occurs if and only if we get $n-1$ points first and then get 2 points. Therefore

$$1-P_n=\frac{1}{2}P_{n-1}.$$

Then we have

$$\begin{aligned}P_n &= 1-\frac{1}{2}P_{n-1}=1-\frac{1}{2}+\frac{1}{4}P_{n-2}=\cdots\\&=1-\frac{1}{2}+\frac{1}{4}-\cdots+(-1)^n\cdot\frac{1}{2^{n-1}}\cdot P_1\\&=1-\frac{1}{2}+\frac{1}{4}-\cdots+\frac{(-1)^n}{2^n}\\&=\frac{1}{3}\left(2+\left(-\frac{1}{2}\right)^n\right).\end{aligned}$$

106. Let the required probability be P_n. Then it is obvious that $P_1=1$, $P_2=\frac{2}{3}$. Assume that $P_k=\frac{2^k}{(k+1)!}$. For $n=k+1$, the maximum number appears at the $(k+1)$th row is

$$\frac{k+1}{\frac{(k+1)(k+2)}{2}}=\frac{2}{k+2}.$$

Then

$$P_{k+1}=\frac{2}{k+2}P_k=\frac{2^{k+1}}{(k+2)!}.$$

Therefore, for any n, we have $P_n=\frac{2^n}{(n+1)!}$.

107. Assume that, after n seconds, the probability that the particle is in place A is P_n, and that it is in B is Q_n. Then

$$P_{n+1} = \frac{1}{2}P_n + Q_n = \frac{1}{2}P_n + (1 - P_n) = 1 - \frac{1}{2}P_n,$$

$$P_{n+1} - \frac{2}{3} = \frac{1}{2}\left(\frac{2}{3} - P_n\right) = \cdots = \left(-\frac{1}{2}\right)^{n+1}\left(P_0 - \frac{2}{3}\right) = \frac{1}{3}\left(-\frac{1}{2}\right)^{n+1},$$

$$P_{n+1} = \frac{2}{3} + \frac{1}{3}\left(-\frac{1}{2}\right)^{n+1}.$$

Therefore $P_n = \frac{2}{3} + \frac{1}{3}\left(-\frac{1}{2}\right)^n$, and $P_{10} = \frac{683}{1\,024}$.

Printed in the United States
By Bookmasters

Printed in the United States
By Bookmasters